BrightRED Revision

Advanced
Higher PHYSICS

Andrew McGuigan

First published in 2009 by:

Bright Red Publishing Ltd
6 Stafford Street
Edinburgh
EH3 7AU

A CIP record for this book is available from the British Library

ISBN 978-1-906736-20-0

With thanks to Ken Vail Graphic Design, Cambridge (layout), Project One Publishing Solutions and Ivor Normand (editorial)

Cover design by Caleb Rutherford at eidetic

Illustrations by Beehive Illustration (Mark Turner), Graham-Cameron Illustration (Graeme Wilson) and Ken Vail Graphic Design

Acknowledgements
Every effort has been made to seek all copyright-holders. If any have been overlooked, then Bright Red Publishing will be delighted to make the necessary arrangements.

The publishers would like to thank the following for the permission to reproduce the following photographs: Meteor impact © NASA (p23); Earth and Moon © NASA/JPL (p24); Saturn with Titan © NASA/JPL/Space Science Institute (p25); Bubble chamber CERN-EX-23296 © CERN Geneva (p55); Aurora Borealis © Joshua Strang (p57); goldfish © Ron Horii (p83).

Printed and bound in Scotland by Scotprint

CONTENTS

INTRODUCTION

AIMS AND STRUCTURE OF THIS BOOK

The aim of this book is to help you revise for your Advanced Higher Physics examination, by presenting the subject arrangements in an attractive and concise format.

Each key sub-topic in the arrangements is presented as a double-page spread, making the book ideal both for revision, or for catching up on key topic areas during your course.

Each double-page spread aims to cover the physics content in a logical and accessible way, and makes full use of graphics and colour illustrations to support your learning. Spreads contain worked examples demonstrating how each Advanced Higher physics equation is used to solve numerical problems, as well as other numerical examples for you to try and build your confidence. Answers to these examples are on pages 92-3.

On each spread you will find 'Don't Forget' pointers, which are there to flag up key points or common mistakes. We have also included 'Internet Links' features, which suggest sites for additional information, interactive simulations or current practical applications of the topic in question. It is worth noting that the text shows examinable derivations where appropriate. Finally, each topic spread finishes with a 'Let's think about this' section, designed to extend and expand your knowledge and interest in the nature of physics and its applications.

Three double page spreads have been allocated to the treatment of uncertainties at Advanced Higher level (see pages 84-9). This treatment can apply to all units in the arrangements. In addition, one double page has been included offering advice on the production of the Investigation Report (pages 90-1).

COURSE STRUCTURE AND ASSESSMENT

The Advanced Higher physics course is divided into four units:

- Mechanics
- Electrical Phenomena
- Wave Phenomena (½ unit)
- Physics Investigation (½ unit)

Students are required to pass an assessment (or "NAB") on completion of each unit. This is marked internally by the teacher. Students should also achieve a pass award for a report on a piece of practical work. This is also marked internally by the teacher and is separate from the investigation report.

The external assessment consists of two parts:

- The completed Investigation Report is sent to SQA for marking, usually during the last week in April. A total of 25 marks is allocated to the Investigation Report.
- A final examination paper of 2½ hours' duration is allocated 100 marks.

The two marks are added together to give a total out of 125, and the course award is graded A, B, C or D depending on the overall mark out of 125.

FORMULAE

Mechanics

$$a = \frac{dv}{dt} = \frac{d^2s}{dt^2}$$

$$v = u + at$$

$$s = ut + \frac{1}{2}at^2$$

$$v^2 = u^2 + 2as$$

$$m = \frac{m_0}{\sqrt{1 - \frac{v^2}{c^2}}}$$

$$E = mc^2$$

$$\omega = \frac{d\theta}{dt}$$

$$\alpha = \frac{d\omega}{dt} = \frac{d^2\theta}{dt^2}$$

$$\omega = \omega_0 + \alpha t$$

$$\theta = \omega_0 t + \frac{1}{2}\alpha t^2$$

$$\omega^2 = \omega_0^2 + 2\alpha\theta$$

$$s = r\theta$$

$$v = r\omega$$

$$a_t = r\alpha$$

$$a_r = \frac{v^2}{r} = r\omega^2$$

$$F = \frac{mv^2}{r} = mr\omega^2$$

$$T = Fr$$

$$T = I\alpha$$

$$L = mvr = mr^2\omega$$

$$L = I\omega$$

$$E_{rot} = \frac{1}{2}I\omega^2$$

$$F = \frac{Gm_1 m_2}{r^2}$$

$$V = -\frac{Gm}{r}$$

$$v = \sqrt{\frac{2Gm}{r}}$$

$$\omega = 2\pi f$$

$$\frac{d^2y}{dt^2} = -\omega^2 y$$

$$y = A\cos\omega t \text{ or } y = A\sin\omega t$$

$$v = \pm\omega\sqrt{(A^2 - y^2)}$$

$$E_k = \frac{1}{2}m\omega^2(A^2 - y^2)$$

$$E_p = \frac{1}{2}m\omega^2 y^2$$

$$\lambda = \frac{h}{p}$$

$$mvr = \frac{nh}{2\pi}$$

point mass
$$I = mr^2$$

rod about centre
$$I = \frac{1}{12}ml^2$$

rod about end
$$I = \frac{1}{3}ml^2$$

disc about centre
$$I = \frac{1}{2}mr^2$$

sphere about centre
$$I = \frac{2}{5}mr^2$$

Electrical phenomena

$$F = \frac{Q_1 Q_2}{4\pi\varepsilon_0 r^2}$$

$$E = \frac{Q}{4\pi\varepsilon_0 r^2}$$

$$V = \frac{Q}{4\pi\varepsilon_0 r}$$

$$F = QE$$

$$V = Ed$$

$$F = IlB\sin\theta$$

$$B = \frac{\mu_0 I}{2\pi r}$$

$$\frac{F}{l} = \frac{\mu_0 I_1 I_2}{2\pi r}$$

$$F = qvB$$

$$\varepsilon = -L\frac{dI}{dt}$$

$$E = \frac{1}{2}LI^2$$

Waves

$$y = A\sin 2\pi\left(ft - \frac{x}{\lambda}\right)$$

$$f = f_s\left(\frac{v}{v \pm v_s}\right)$$

$$f = f_s\left(\frac{v \pm v_0}{v}\right)$$

$$\Phi = \frac{2\pi x}{\lambda}$$

optical path difference $= m\lambda$

optical path difference $= (m + \frac{1}{2})\lambda$

$$\Delta x = \frac{\lambda l}{2d}$$

$$d = \frac{\lambda}{4n}$$

$$\Delta x = \frac{\lambda D}{d}$$

$$n = \tan i_p$$

$$\frac{\Delta X}{X} = \sqrt{\left(\frac{\Delta Y}{Y}\right)^2 + \left(\frac{\Delta Z}{Z}\right)^2}$$

$$\Delta X = \sqrt{\Delta Y^2 + \Delta Z^2}$$

$$\Delta m = \frac{m_1 - m_2}{2\sqrt{(n - 2)}}$$

$$\Delta c = \frac{c_1 - c_2}{2\sqrt{(n - 2)}}$$

Please also refer to the Data Booklet for Advanced Higher Physics, which is available as a download from www.sqa.org.uk/files_ccc/Physics_data_booklet.pdf

KINEMATIC RELATIONSHIPS AND RELATIVISTIC MOTION: 1

In earlier studies of physics, we used three equations of motion for objects moving in a straight line with constant acceleration:

$$v = u + at$$

$$s = ut + \frac{1}{2}at^2$$

$$v^2 = u^2 + 2as \quad \text{where the symbols have their usual meanings.}$$

In Advanced Higher Physics, we begin by deriving these equations using calculus.

DERIVATION OF KINEMATIC RELATIONSHIPS

Velocity **v** is defined as the rate of change of displacement **s** or $v = \dfrac{ds}{dt}$

Acceleration **a** is defined as the rate of change of velocity **v** or $a = \dfrac{dv}{dt}$

Combining, we get $\mathbf{a} = \dfrac{dv}{dt} = \dfrac{d}{dt}\left(\dfrac{ds}{dt}\right) = \dfrac{d^2s}{dt^2}$

Deriving $v = u + at$

$$\frac{dv}{dt} = a$$

$$dv = a\,dt$$

$$\int_u^v dv = \int_0^t a\,dt$$

$$[v]_u^v = [at]_0^t$$

$$v - u = at$$

$$v = u + at$$

Deriving $s = ut + \frac{1}{2}at^2$

$$\frac{ds}{dt} = v$$

$$ds = v\,dt$$

$$\int_0^s ds = \int_0^t v\,dt = \int_0^t (u + at)dt = \int_0^t u\,dt + at\,dt$$

$$[s]_0^s = [ut + \tfrac{1}{2}at^2]_0^t$$

$$s = ut + \tfrac{1}{2}at^2$$

Official SQA Past Papers Adva

1. (a) An object moves with constant acceleration a. At time $t = 0$ its displacement s is zero. The velocity v of the object is given by $v = u + at$. Derive the equation

$$s = ut + \tfrac{1}{2}at^2$$

where the symbols have their usual meanings.

Deriving $v^2 = u^2 + 2as$

Having first derived $v = u + at$ and $s = ut + \frac{1}{2}at^2$, use calculus and rearrange:

$$v = u + at \quad \text{to get} \quad t = \frac{v - u}{a}$$

Substitute this into $s = ut + \frac{1}{2}at^2$

$$s = u\frac{(v - u)}{a} + \frac{1}{2}a\frac{(v - u)^2}{a^2}$$

$$2as = 2uv - 2u^2 + v^2 - 2uv + u^2$$

$$v^2 = u^2 + 2as$$

contd

DERIVATION OF KINEMATIC RELATIONSHIPS contd

Worked example

An arrow is fired vertically into the air, and its vertical displacement s is given by

$$s = 44 \cdot 1t - 4 \cdot 9t^2 \text{ metres} \qquad t \text{ is in seconds}$$

a Find an expression for the velocity of the arrow.

$$v = \frac{ds}{dt} = 44 \cdot 1 - 9 \cdot 8t$$

b Find the acceleration of the arrow.

$$a = \frac{dv}{dt} = -9 \cdot 8 \text{ ms}^{-2}$$

c Calculate the initial velocity of the arrow.

$$v = 44 \cdot 1 - 9 \cdot 8t \text{ when } t = 0$$
$$= 44 \cdot 1 - 0$$
$$v = 44 \cdot 1 \text{ ms}^{-1}$$

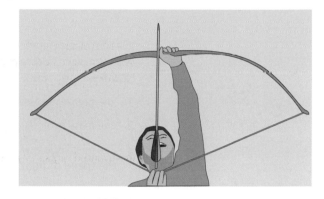

d Calculate the maximum height reached by the arrow.

Max height when $v = 0$: $\quad 44 \cdot 1 - 9 \cdot 8t = 0, \quad t = \dfrac{44 \cdot 1}{9 \cdot 8} = 4 \cdot 5 \text{ s}$

$$s = 44 \cdot 1 \times 4 \cdot 5 - 4 \cdot 9 \times 4 \cdot 5^2 = 99 \cdot 2 \text{ m}$$

Example 1

The displacement **s** of an object in metres is given by

$$s = 15 + 20t^2 - 30t^4 \qquad (\text{time } t \text{ is in seconds})$$

a Find an expression for the velocity as a function of time.

b Find an expression for the acceleration as a function of time.

c What was the initial displacement of the object?

d What was the initial velocity of the object?

e What was the initial acceleration of the object?

f At what times is the velocity zero?

Example 2

The displacement **s** of a rocket from its launch pad is given by:

$$s = t^3 + 6t^2 \text{ metres, for } 0 \le t \le 20 \text{ seconds.}$$

a Find an expression for the velocity as a function of time.

b Find an expression for the acceleration as a function of time.

c Calculate the speed of the rocket after 12 seconds.

d The fuel burned for 20 seconds. How far had the rocket travelled in that time?

Visit http://hypertextbook.com/physics/mechanics/motion-equations for many interesting examples with solutions.

LET'S THINK ABOUT THIS

1 In Example 2, how long will the rocket take to reach Mach 1, the speed of sound (340 ms^{-1})? You will need to solve a quadratic equation to do this.

2 The first two derivations make regular appearances in prelims, NABs and final exams.

KINEMATIC RELATIONSHIPS AND RELATIVISTIC MOTION: 2

In 1905, Albert Einstein put forward his **Special Theory of Relativity**, which proposed that objects moving at speeds near to the speed of light ($3 \times 10^8\,\text{ms}^{-1}$) did not obey the laws of Newtonian mechanics. Three of Einstein's proposals were:

1 The **greatest possible speed** of an object is **always less** than the **speed of light** in a vacuum.

2 The mass *m* of a moving object, called the **relativistic mass**, increases as the speed increases.

3 The relativistic energy *E* of a moving object is given by the relationship $E = mc^2$ where *c* is the **speed of light**.

RELATIVISTIC MASS

The mass of a stationary object is called the **rest mass** and is represented by the symbol m_0. If the mass is moving at a speed *v* then its relativistic mass *m* is given by:

$$m = \frac{m_0}{\sqrt{1 - \dfrac{v^2}{c^2}}}$$

Worked example

Calculate the relativistic mass of an electron travelling at $1\cdot8 \times 10^8\,\text{ms}^{-1}$. The rest mass of an electron is $9\cdot11 \times 10^{-31}\,\text{kg}$.

$$m = \frac{m_0}{\sqrt{1 - \dfrac{v^2}{c^2}}}$$

$$= \frac{9\cdot11 \times 10^{-31}}{\sqrt{1 - \dfrac{1\cdot8 \times 10^{8^2}}{3 \times 10^{8^2}}}}$$

$$= \frac{9\cdot11 \times 10^{-31}}{\sqrt{1 - 0\cdot36}}$$

$$= 1\cdot14 \times 10^{-30}\,\text{kg}$$

This represents a 25% increase in mass when the electron travels at $1\cdot8 \times 10^8\,\text{ms}^{-1}$ or 60% of the speed of light.

Example 1

Calculate the relativistic mass of a neutron travelling at $9\cdot5 \times 10^7\,\text{ms}^{-1}$. The rest mass of a neutron is $1\cdot675 \times 10^{-27}\,\text{kg}$.

Example 2

The rest mass of a proton is $1\cdot673 \times 10^{-27}\,\text{kg}$. Calculate the speed of the proton when its relativistic mass is $2\cdot155 \times 10^{-27}\,\text{kg}$.

RELATIVISTIC ENERGY E = mc²

Einstein's famous equation $E = mc^2$ states that the **relativistic energy *E*** of an object is equal to the **relativistic mass** of the object **times** the **speed of light squared**.

$$E = mc^2 = \frac{m_0}{\sqrt{1 - \dfrac{v^2}{c^2}}}c^2$$

This equation suggests that **mass** and **energy** are **interchangeable**. Mass can be turned wholly or partially into energy and vice versa.

contd

RELATIVISTIC ENERGY $E = mc^2$ contd

Mass turned into energy

When a uranium nucleus undergoes fission by a neutron, some of the total mass of the uranium and neutron is turned into energy. The mass of the two fission products and the three neutrons is less than the uranium nucleus and the first neutron. The missing mass has been turned into energy.

Energy turned into mass

Under certain conditions, **gamma rays** can combine to form **two particles** – an **electron** and **positron**. (A positron is a positive electron.) The gamma rays have no mass, and their energy is turned into mass.

(The reverse process is also possible where an electron can combine with a positron to create gamma rays. This is mass being turned into energy.)

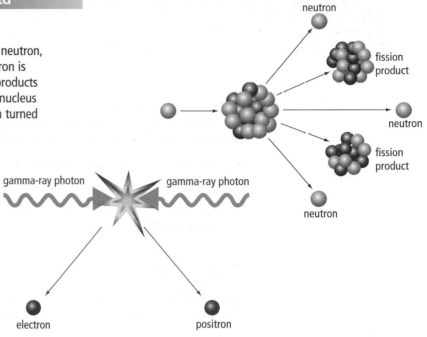

Worked example

An electron has a speed of $2 \cdot 0 \times 10^8\,ms^{-1}$. Calculate the relativistic energy of the electron. The rest mass of an electron is $9 \cdot 11 \times 10^{-31}\,kg$.

$$E = mc^2 = \frac{m_0}{\sqrt{1 - \frac{v^2}{c^2}}}\,c^2 = \frac{9 \cdot 11 \times 10^{-31}}{\sqrt{1 - \frac{2 \times 10^{8^2}}{3 \times 10^{8^2}}}} \cdot (3 \times 10^8)^2 = 1 \cdot 22 \times 10^{-30} \times (9 \times 10^{16}) = 1 \cdot 1 \times 10^{-13}\,J$$

Example 3

A proton has a speed of $2 \cdot 3 \times 10^8\,ms^{-1}$. Calculate the relativistic energy of the proton. The rest mass of a proton is $1 \cdot 673 \times 10^{-27}\,kg$.

Example 4

The relativistic energy of a neutron is $1 \cdot 8 \times 10^{-10}\,J$. Calculate the speed of the neutron. The rest mass of the neutron is $1 \cdot 675 \times 10^{-27}\,kg$.

Visit http://www.btinternet.com/~j.doyle/SR/Emc2/Basics.htm#Introduction for more on $E = mc^2$.

LET'S THINK ABOUT THIS

1 Science-fiction writers often describe spacecraft having speeds greater than the speed of light. Use the relativistic mass equation to show that it is mathematically impossible for a spacecraft to have a speed greater than $3 \times 10^8\,ms^{-1}$.

2 Relativistic effects only become noticeable when the speed of the moving object is greater than 10% of the speed of light. Show that the relativistic mass is only 0·5% greater than the rest mass when the speed is 10% of the speed of light.

3 Physicists often use the shorthand symbol γ (called the **Lorentz factor**) to represent $\frac{1}{\sqrt{1 - \frac{v^2}{c^2}}}$ so the relativistic mass equation becomes a less cluttered $m = \gamma m_0$. (For interest only, and may help understanding if you come across some complicated-looking relativistic mathematics.)

MECHANICS

ANGULAR MOTION

You have already studied linear motion in Standard Grade and Higher Physics. In the Advanced Higher, this work is extended to cover circular motion in which objects move along circular paths rather than linear paths.

ANGULAR VELOCITY

DON'T FORGET

The abbreviation is rad s^{-1}, not rads s^{-1}

A useful quantity in circular motion is **angular velocity** ω which is a measure of the **angle swept out per second**.

Consider an object moving at a steady linear speed v in a circle with centre O. The angle θ is swept out by the radius as the object moves from A to B in a time of t seconds.

θ is measured in **radians**, ω is measured in **radians per second** or **rads^{-1}**.

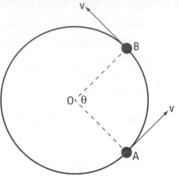

Worked example

A wind generator makes 6 complete revolutions in a time of 10 seconds. Calculate the angular velocity of the wind generator.

1 complete revolution $= 2\pi$ radians

$$\omega = \frac{\theta}{t} = \frac{6 \times 2\pi}{10} = 3 \cdot 8 \, \text{rads}^{-1}$$

There are several other ways in which ω can be calculated.

1 The number of revolutions per minute (rpm) could be given, e.g. an electric drill rotates at 750 rpm.

$$\omega = \frac{\theta}{t} = \frac{750 \times 2\pi}{60} = 78 \cdot 5 \, \text{rads}^{-1}$$

2 The time for one complete revolution is called the **period**, T.

$$\omega = \frac{\theta}{t} = \frac{2\pi}{T}$$

3 The frequency f of rotation is related to the period T by $f = \frac{1}{T}$, so substituting gives

$$\omega = 2\pi f$$

4 Differentiation can be used to calculate ω if an expression for θ as a function of time is given:

$$\omega = \frac{d\theta}{dt}$$

Example 1

A microwave tray rotates with a period of 4 seconds. Calculate its angular velocity.

Example 2

Calculate the angular velocity of the Earth as it rotates about its polar axis.

Example 3

Mars rotates about its polar axis with an angular velocity of $7 \cdot 1 \times 10^{-5}$ rads^{-1} and rotates around the Sun with an angular velocity of $1 \cdot 06 \times 10^{-7}$ rads^{-1}. Calculate the length of a Martian day and a Martian year.

Example 4

How many radians does the Earth turn through in 1 year?

ANGULAR VELOCITY AND TANGENTIAL SPEED

Objects moving in a circle also have a **linear speed** in the **direction of the tangent**. This **linear** or **tangential speed v** is related to the **angular speed** ω. Consider one complete orbit in a circle of **radius r**.

$$v = \frac{distance}{time} = \frac{circumference}{time} = \frac{2\pi r}{T} = \frac{2\pi}{T}r = \omega r$$

$$v = \omega r$$

Worked example

A ceiling fan rotates at 150 rpm. Each blade of the fan is 45 cm long.

a Calculate the angular velocity of the fan.

$$\omega = \frac{\theta}{t} = \frac{150 \times 2\pi}{60} = 15\cdot7\,\text{rads}^{-1}$$

b Calculate the speed of the tip of each blade.

$$v = \omega r = 15\cdot7 \times 0\cdot45 = 7\cdot1\,\text{ms}^{-1}$$

c Calculate the speed of a point halfway along a blade.

$$v = \omega r = 15\cdot7 \times 0\cdot225 = 3\cdot5\,\text{ms}^{-1}$$

Note that **points closer to the axis** of rotation will have **lower tangential speeds** than points further from the axis. The *r* in **v** = ω*r* is the **distance from the axis of rotation** to the point whose tangential speed is being calculated.

Example 5

Assuming the Earth to be spherical with radius $6\cdot4 \times 10^6$ m, calculate:

a the tangential speed of a person standing on the Equator.

b the tangential speed of a person standing in central Scotland, which has a latitude of 56°.

Example 6

Calculate the linear speed of the tip of the minute hand of a watch. The minute hand is 1·2 cm long and moves continuously.

Example 7

Calculate the linear speed of the Earth as it orbits the Sun. The Earth's mean orbit radius is $1\cdot5 \times 10^{11}$ m.

> Visit http://www.dctech.com/physics/help/problems.php?problem=circular-velocity for more information and examples of angular velocity

RADIAN MEASURE

An angle in radians is defined as $\theta = \dfrac{arc\ length}{radius} = \dfrac{s}{r}$. This relationship is given in the physics data booklet as **s = rθ**.

LET'S THINK ABOUT THIS

A star is 3000 light years from the centre of its rotating constellation and moves with a linear speed of 113 kms⁻¹. Show that it makes one complete orbit every 50 million years.

ANGULAR ACCELERATION

When the **angular velocity of a rotating object changes**, it is said to have an **angular acceleration**. Angular acceleration has the symbol α and unit **rads^{-2}**. A constant angular acceleration of 3 rads^{-2} means the angular velocity increases by 3 rads^{-1} every second.

Angular acceleration can be described mathematically:

$$\alpha = \frac{d\omega}{dt} \qquad \text{angular acceleration is the rate of change of angular velocity}$$

$$\alpha = \frac{d^2\theta}{dt^2} \qquad \text{since } \omega = \frac{d\theta}{dt} \text{ then } \alpha \text{ is the second derivative of } \theta(t).$$

These two formulae will give the **instantaneous angular acceleration** at time t.

Worked example

The angular displacement of a particle performing circular motion is given by:

$$\theta = 4t^2 + 3t + 2 \text{ radians.}$$

Calculate the particle's angular velocity and angular acceleration after 5 s.

$$\omega = \frac{d\theta}{dt} = 8t + 3 = (8 \times 5) + 3 = 43 \text{ rads}^{-1}$$

$$\alpha = \frac{d\omega}{dt}$$
$$= 8 \text{ rads}^{-2} \; (= \text{ a constant acceleration independent of time } t)$$

Example 1

The angular displacements of two objects A and B are given by:

object A: $\qquad \theta = -2 - 0.6t + 0.35t^2$

object B: $\qquad \theta = 0.5t + 0.15t^3$

Calculate the angular acceleration of each object after 4 seconds.

Constant angular acceleration

If the angular acceleration is **constant**, then the following expression can be used to calculate α.

$$\alpha = \frac{\omega - \omega_0}{t} \qquad \text{where } \omega_0 = \text{initial angular velocity (rads}^{-1})$$
$$\omega = \textbf{final angular velocity (rads}^{-1}\textbf{)} \text{ after time } t \, (s)$$

Worked example

During take-off, the main rotor blades of a helicopter increase their rate of rotation uniformly from rest to 450 rpm in 8 seconds. Calculate the angular acceleration of the rotor blades.

$$450 \text{ rpm} = \frac{450 \times 2 \times 3.14}{60}$$
$$= 47.1 \text{ rads}^{-1}$$

$$\alpha = \frac{\omega - \omega_0}{t}$$
$$= \frac{47.1 - 0}{8}$$
$$= 5.9 \text{ rads}^{-2}$$

Note that the tip of the blade will have an increasing **linear** speed as the angular speed increases, giving rise to the concept of tangential acceleration a_t.

If the helicopter blade is 6.5 m long, then the tangential acceleration of the tip of the blade can be calculated:

$$a_t = \alpha r = 5.9 \times 6.5 = 38 \text{ ms}^{-2}.$$

contd

ANGULAR ACCELERATION contd

Kinematic relationships for constant angular acceleration

The following equations for constant angular acceleration are exactly analogous to those for constant linear acceleration which were derived on page 4:

$$\omega = \omega_0 + \alpha_t$$ where ω_0 = initial angular velocity (rads^{-1})

$$\theta = \omega_0 t + \frac{1}{2}\alpha t^2$$ ω = final angular velocity (rads^{-1})

$$\omega^2 = \omega_0^2 + 2\alpha\theta$$ α = constant angular acceleration (rads^{-2})

θ = angular displacement (rad)

t = time taken (s)

Worked example

A bicycle wheel rotating at 4 revolutions per second comes to rest uniformly in a time of 2 minutes. Calculate the angular acceleration of the wheel during this time.

$$\omega_0 = \frac{\theta}{t} = \frac{4 \times 2 \times 3.14}{1} = 25.1 \text{ rads}^{-1}.$$

$$\alpha = \frac{\omega - \omega_0}{t} = \frac{0 - 25.1}{120} = -0.21 \text{ rads}^{-2}.$$

How many revolutions does the wheel make as it decelerates?

$$\theta = \omega_0 t + \frac{1}{2}\alpha t^2 = 25.1 \times 120 + 0.5 \times (-0.21) \times 120^2$$

$$= 1500 \text{ rad} = \frac{1500}{2 \times 3.14} = 238.9 \text{ revolutions.}$$

Example 2

A spin dryer rotating at 800 rpm makes 95 complete revolutions as it slows down uniformly and stops. Calculate the time taken by the spin dryer to decelerate and stop.

Example 3

A spinning disc accelerates uniformly from 30 rpm to 72 rpm in a time of 4 seconds. Calculate the number of revolutions made by the disc in this time.

Example 4

A lawnmower blade rotating at 300 rpm is switched off. It comes to rest after making 4 complete revolutions. Calculate the time taken by the blade to stop.

Example 5

The graph shows how the angular velocity of a rotating disc varies with time.

a Calculate the initial acceleration and final acceleration of the disc.

b Calculate the total angular displacement of the disc.

c How many revolutions does the disc make?

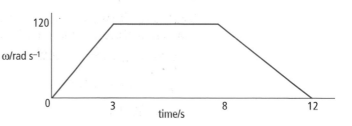

LET'S THINK ABOUT THIS

1 Take care with decelerations. The angular acceleration will have a negative value, and this must be included when substituting.

2 Avoid confusing linear and angular expressions which are similar. Writing

$s = \omega_0 t + \frac{1}{2}\alpha^2$ instead of $\theta = \omega_0 t + \frac{1}{2}\alpha t^2$ could be penalised following the

introduction of the data booklet.

MECHANICS

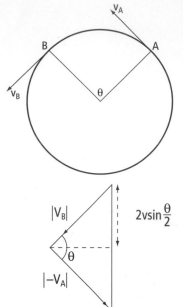

CENTRIPETAL ACCELERATION

When an object moves in a circle with a **steady speed v**, its direction is continually changing. Velocity is a **vector** requiring both **magnitude** and **direction**, so the **velocity** of the object will be **changing** even if the **speed is constant**. The object must be **accelerating** as its **velocity is changing**.

Derivation of $a = \dfrac{v^2}{r} = \omega^2 r$ (difficult!)

The change of velocity from A to B is found using a vector addition diagram.

$$V_B - V_A = \diagup - \diagdown = \diagup + \diagdown = V_B \diagdown = \left| 2v\sin\frac{\theta}{2} \right.$$

| see triangle diagram for the maths

Notice the change in velocity from A to B is directed (down) towards the centre of the circle.

The time to travel from A to B $= \dfrac{arc\ length\ AB}{v} = \dfrac{r\theta}{v}$

acceleration towards centre $= \dfrac{\Delta v}{\Delta t} = \dfrac{2v\sin\frac{\theta}{2}}{\frac{r\theta}{v}}$ (note: $\sin x = x$ for small values of x.)

$$= \dfrac{2v\frac{\theta}{2}}{\frac{r\theta}{v}}$$

$$= \dfrac{v^2}{r} = \dfrac{\omega^2 r^2}{r}$$

$$a = \omega^2 r.$$

This acceleration towards the centre of rotation is called the **centripetal acceleration** or **radial acceleration**.

Worked example

A radio-controlled model aircraft moves in a circle of radius 7·5 m. Calculate the centripetal acceleration of the aircraft when (a) it travels at a steady speed of 8·2 ms⁻¹ (b) the aircraft takes 4·5 seconds to make 1 complete circuit.

a $a = \dfrac{v^2}{r} = \dfrac{8\cdot2^2}{7\cdot5} = 9\,\text{ms}^{-2}$

b $\omega = \dfrac{2\pi}{t} = \dfrac{2 \times 3\cdot14}{4.5} = 1\cdot4\,\text{rads}^{-1}$

$a = \omega^2 r = 1\cdot4^2 \times 7\cdot5 = 14\cdot7\,\text{ms}^{-2}$

We have now met three separate accelerations associated with circular motion, and it is important to understand their differences:

Angular acceleration	$\alpha = \dfrac{\omega - \omega_0}{t}$	unit is rads⁻²	increasing/decreasing ω (and v)
Tangential acceleration	$a_t = \alpha r$	unit is ms⁻²	increasing/decreasing v (and ω)
Centripetal acceleration	$a = \dfrac{v^2}{r} = \omega^2 r$	unit is ms⁻²	ω and v constant (for constant a)

Example 1

An object moves in a circle with radius 4·1 m and an angular velocity 2·6 rads⁻¹. The angular velocity increases uniformly to 3·8 rads⁻¹ in a time of 3·5 s. Calculate:

a the angular acceleration.

b the tangential acceleration.

c the centripetal acceleration of the object **(i)** at the beginning **(ii)** after 3·5 seconds.

CENTRIPETAL FORCE

Objects moving in a circular path will **accelerate towards the centre of the circle**. Newton's second law tells us there must be an **unbalanced force** causing this acceleration. This force is called the **centripetal force** or **central force**.

$$F = ma = m\frac{v^2}{r} = m\omega^2 r$$

Examples of centripetal forces include:

- mass rotating horizontally on the end of a string
- spin dryer
- aircraft turning
- conical pendulum.

Mass rotating horizontally on the end of a string

The tension in the string provides the centripetal force on the mass: $T = m\dfrac{v^2}{r}$

If the string breaks, the centripetal force will be removed and the mass will then move in a straight line, not in a circle (Newton I). A centripetal force is required to maintain circular motion.

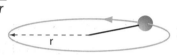

Example 2

Calculate the centripetal force on a mass of 0·4 kg making 3 revolutions per second in a circle of radius 0·5 m.

Spin dryer

The wet clothes will move in a circle due to the reaction force provided by the drum on the clothes. The water over the holes in the drum will not be in contact with the drum and will not experience a centripetal force. This water will move tangentially in a straight line. It is a common misconception that the water will fly out radially due to some 'outward force'. This is not correct.

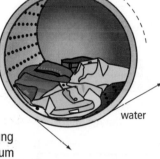

water

Example 3

A damp sock of mass 40 g experiences a centripetal force of 78·1 N from the drum of a washing machine on spin cycle. The radius of the drum is 22 cm. Calculate the angular speed of the drum in rpm on the spin cycle.

Aircraft turning (banking)

An aircraft in level flight has its weight balanced by the lift force **L**. When banking at angle θ, the horizontal component of **L** provides the centripetal force.

$$L\sin\theta = m\frac{v^2}{r}$$

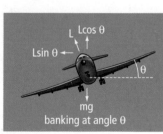

L

mg
level flight

L Lcos θ

Lsin θ

θ

mg
banking at angle θ

Notice that **mg > Lcosθ** (vertical forces), so the aircraft may lose height when banking unless the lift **L** is increased.

Conical pendulum

A conical pendulum consists of a mass **m** on the end of a string. The mass travels in a horizontal circle as shown. The centripetal force is provided by the horizontal component of the tension **T** in the string.

$$T\sin\theta = m\frac{v^2}{r} = m\omega^2 r$$

$$T\cos\theta = mg \qquad \text{divide first equation by the second}$$

$$\frac{T\sin\theta}{T\cos\theta} = \frac{m\frac{v^2}{r}}{mg} = \frac{v^2}{rg}$$

$$\tan\theta = \frac{v^2}{rg} = \frac{r\omega^2}{g} \qquad \text{Also } \sin\theta = \frac{L}{r} \text{ where L is the length of the string.}$$

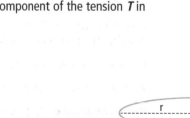

θ

T

Tcosθ

r

Tsinθ

mg

Visit http://www.glenbrook.K12.il.US/GBSSCI/PHYS/Class/circles/u612a.html to see how **friction provides the centripetal force** for a car travelling round a bend in the road.

ROTATIONAL DYNAMICS

MOMENT OF INERTIA

The mass of a rotating object will be an important consideration when calculating the object's kinetic energy. The distribution of this mass about the axis of rotation is an equally important factor.

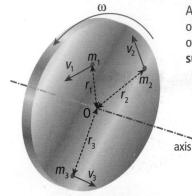

An object rotates with **uniform angular velocity** ω about an axis through **O**. Consider the object as a series of individual point masses m_1, m_2, m_3, ... at different distances from the axis of rotation as shown in the diagram. The **kinetic energy** E_k of the rotating object will be the **sum of the kinetic energy of each point mass**.

$$E_k = \tfrac{1}{2}m_1v_1^2 + \tfrac{1}{2}m_1v_1^2 + \tfrac{1}{2}m_1v_1^2 \ldots$$
$$= \tfrac{1}{2}m_1\omega^2r_1^2 + \tfrac{1}{2}m_2\omega^2r_2^2 + \tfrac{1}{2}m_3\omega^2r_3^2 \ldots$$
$$= \tfrac{1}{2}(m_1r_1^2 + m_2r_2^2 + m_3r_3^2 + \ldots)\omega^2$$
$$= \tfrac{1}{2}(\Sigma mr^2)\omega^2$$
$$= \tfrac{1}{2}I\omega^2$$

where I is the sum of all the mr^2 values for all the particles in the object.

I is called the **moment of inertia** of the object and is a measure of an object's **resistance to change in its rotation rate**. I has the unit **kgm²**.

The moment of inertia I of a rotating object depends on the **mass** of the object and the **distribution of the mass** about the axis of rotation and can be calculated using integration (not in the AH Physics syllabus).

Expressions for the moment of inertia for five different objects rotating about certain axes are given.

DON'T FORGET

These five expressions are given in the AH data booklet.

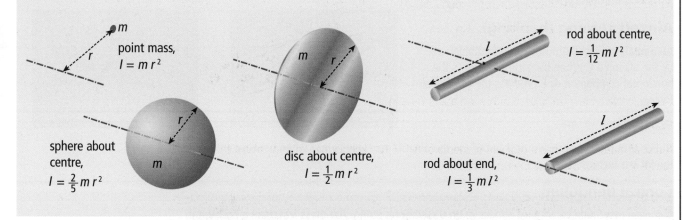

point mass,
$I = mr^2$

sphere about centre,
$I = \tfrac{2}{5}mr^2$

disc about centre,
$I = \tfrac{1}{2}mr^2$

rod about centre,
$I = \tfrac{1}{12}ml^2$

rod about end,
$I = \tfrac{1}{3}ml^2$

Worked example

A disc of mass 0·24 kg and radius 15 cm rotates about the axis through its centre.

Calculate the moment of inertia of this disc.

$$I = \tfrac{1}{2}mr^2 = 0{\cdot}5 \times 0{\cdot}24 \times 0{\cdot}15^2 = 2{\cdot}7 \times 10^{-3}\,\text{kgm}^2$$

Worked example

A rod of mass 0·21 kg and length 0·5 m is fixed to a disc of mass 0·12 kg and radius 0·18 m as shown. The rod–disc combination rotates about the axis at 15 rads⁻¹.

Calculate the kinetic energy of the disc–rod combination.

contd

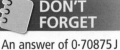

MOMENT OF INERTIA contd

First, calculate the moment of inertia of the rod–disc combination.

$$I_{combination} = I_{disc} + I_{rod} = 0{\cdot}5 \times 0{\cdot}12 \times 0{\cdot}18^2 + \frac{1}{12} \times 0{\cdot}21 \times 0{\cdot}5^2$$
$$= 6{\cdot}3 \times 10^{-3}\ kgm^2$$
$$E_k = \frac{1}{2}I\omega^2 = 0{\cdot}5 \times 6{\cdot}3 \times 10^{-3} \times 15^2 = 0{\cdot}71\ J$$

Example 1

A sphere of mass 0·80 kg and radius 0·25 m rotates about an axis through its centre. Calculate the moment of inertia of the sphere about this axis.

Example 2

The Earth has a mass of $6{\cdot}0 \times 10^{24}$ kg and a radius of $6{\cdot}4 \times 10^6$ m.

a Calculate the angular velocity of the Earth in $rads^{-1}$.

b Calculate the moment of inertia of the Earth, assuming the Earth has uniform density.

c Calculate the kinetic energy of the rotating Earth.

Example 3

A compact disc (CD) has a mass of 14 g and radius of 58 mm.

a Show that the moment of inertia of the CD is approximately $2{\cdot}35 \times 10^{-5}$ kgm² when operating normally.

b Show that the kinetic energy of the CD is approximately $5{\cdot}9 \times 10^{-3}$ J when the angular velocity of the CD is $22{\cdot}4\ rads^{-1}$.

c In practice, the CD has a central hole for fitting into the CD player. Will the moment of inertia of the CD be greater or less than $2{\cdot}35 \times 10^{-3}$ kgm²? Justify your answer.

Example 4

A point mass of 0·15 kg sits on a disc of mass 0·26 kg and radius 0·22 m. The mass is 0·14 m from the centre of the disc. The disc and mass rotate at $8{\cdot}5\ rads^{-1}$ about an axis through the centre of the disc.

Calculate the kinetic energy of the disc and mass.

axis

0.14 m

0.22 m

Example 5

A rod of length 0·60 m and mass 0·15 kg rotates about an axis through one of its ends.

Calculate the moment of inertia of the rod about this axis.

Example 6

A disc of mass 0·72 kg has a moment of inertia of $8{\cdot}1 \times 10^{-3}$ kgm² when rotating about the centre of the disc. Calculate the radius of the disc.

DON'T FORGET

An answer of 0·70875 J has too many significant figures.

DON'T FORGET

You must calculate I before using $E_k = \frac{1}{2}I\omega^2$

Look up http://hyperphysics.phy-astr.gsu.edu/hbase/mi.html to reinforce your knowledge of moment of inertia.

LET'S THINK ABOUT THIS

1 Revisit Example 2. In reality, the Earth is not a uniform sphere. The core of the Earth is denser than the region nearer the Earth's crust. What effect will this have on the moment of inertia of the Earth and hence the rotational kinetic energy?

2 $I_{rod\ end} > I_{rod\ centre}$ as more of the rod mass is at a greater distance from the axis of rotation.

3 The size of the radius of a rotating disc has a greater overall effect on the value of I than the mass of the disc. If the mass of the disc doubles, then I doubles. If the radius doubles, then I increases four times because the radius is squared in the expression for I_{disc}.

TORQUE

Torque *T* is the **turning effect** of a **force** on a **rotating object**. The disc can be made to turn about its **axis** by applying a **force *F*** at a **perpendicular distance *r*** from the axis of rotation. The turning effect on the disc will be less if distance *r* is reduced.

torque = applied force × perpendicular distance between direction of force and axis of rotation

$$T = F \times r \qquad \text{The unit of torque is Nm}$$

Torque is sometimes called the **moment of a force**.

Worked example

A force of 65 N is applied tangentially to a disc of radius 0·20 m. The disc can rotate about an axis through its centre. Calculate the applied torque.

$$T = F \times r = 65 \times 0.2 = 13\,\text{Nm}.$$

Example 1

A torque of 26 Nm is applied to a nut using a spanner. Calculate the force exerted by the operator.

Example 2

A 57 N force is applied to a spanner at an angle of 70° as shown. The spanner is held 28 cm from the nut. Calculate the torque applied to the nut.

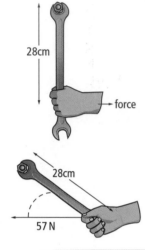

28cm

28cm

57 N

TORQUE AND ANGULAR ACCELERATION

> **DON'T FORGET**
>
> The symbol *T* in $T = I\alpha$ is the **unbalanced torque**.

An **unbalanced torque** on an object will produce an **angular acceleration** about an axis of rotation. The **angular acceleration** produced by the **unbalanced torque** depends on the **moment of inertia** of the object.

The relationship linking **torque *T*** and **angular acceleration** α is:

$$T = I\alpha \quad \text{where } I \text{ is the moment of inertia of the object about the axis.}$$

This is the **rotational analogue** to Newton's second law $F = ma$

If a rotating object is subjected to friction, then a **frictional torque** will **oppose** the **angular acceleration**.

unbalanced torque = applied torque – frictional torque

You are already familiar with finding the unbalanced force on an object. Unbalanced torque is found in a similar way, as the following diagrams show.

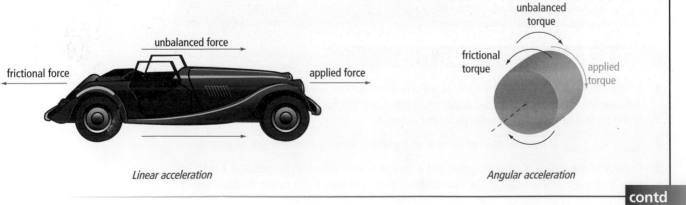

Linear acceleration

Angular acceleration

contd

TORQUE AND ANGULAR ACCELERATION contd

Worked example

A force of 15 N is applied to a string wrapped round the circumference of a disc of radius 0·30 m. The disc has a mass of 0·40 kg and accelerates at 35 rads⁻² about the axis. Calculate:

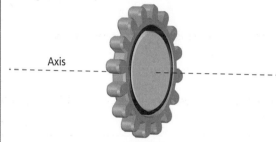

a = 35 rads⁻²

axis

15 N

a the torque applied to the disc (by the string).
Applied torque $= F \times r = 15 \times 0.3 = 4.5\,\text{Nm}$

b the unbalanced torque on the disc.
Calculate I first:
$I = \frac{1}{2}mr^2 = 0.5 \times 0.4 \times 0.3^2 = 1.8 \times 10^{-2}\,\text{kgm}^2$
Unbalanced torque $T = I\alpha = 1.8 \times 10^{-2} \times 35 = 0.63\,\text{Nm}$

c the frictional torque acting on the disc.
Unbalanced torque $=$ applied torque $-$ frictional torque
Frictional torque $=$ applied torque $-$ unbalanced torque $= 4.5 - 0.63 = 3.87\,\text{Nm} = 3.9\,\text{Nm}.$

Example 3

Axis

A flywheel of moment of inertia $7.0 \times 10^{-3}\,\text{kgm}^2$ is accelerated from rest to 65 rads⁻¹. Calculate the unbalanced torque on the flywheel:

a if the acceleration takes 4·0 s

b if the acceleration takes 12 complete revolutions.

⚙ LET'S THINK ABOUT THIS

A bench grinder of moment of inertia 0·48 kgm² rotates at a constant speed of 600 rpm. The grinding wheel is driven by a 500 W electric motor. It can be shown that power
$P =$ Torque $\times \omega$ for objects rotating at steady speeds.

a Show that the frictional torque at steady speed is 8 Nm.

b The motor is now switched off. Show that the wheel takes 3·8 s to come to rest, assuming the frictional torque remains constant. (α is negative, so your calculation for t must include the minus sign.)

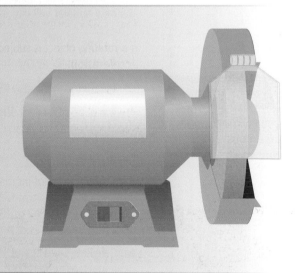

ANGULAR MOMENTUM

The **angular momentum** of a rotating object is defined as:

$$L = I\omega$$

The **unit of angular momentum** is **kgm²rads⁻¹** or **kgm²s⁻¹**.

The unit of angular momentum is $\text{kgm}^2\text{rads}^{-1}$ or $\text{kgm}^2\text{s}^{-1}$.

This is the **rotational analogue** to **linear momentum** $p = mv$ where I replaces m and ω replaces v.

Worked example

A uniform disc of mass 0·55 kg and radius 0·20 m rotates at 32 rads⁻¹. Calculate the angular momentum of the disc.

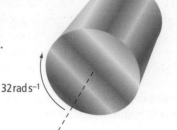

$$I_{disc} = \tfrac{1}{2}mr^2 = 0.5 \times 0.55 \times 0.2^2$$
$$= 1.1 \times 10^{-2}\,\text{kgm}^2$$
$$L = I\omega = 1.1 \times 10^{-2} \times 32 = 0.35\,\text{kgm}^2\text{s}^{-1}$$

32 rads⁻¹

Single particle moving in a circle

A single particle of mass m moving in a circle of radius r with angular velocity ω has an angular momentum as follows.

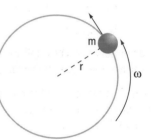

$$L = I\omega = mr^2\omega \text{ substituting } v = \omega r$$
$$L = mr^2\omega = mr^2\,\frac{v}{r} = mrv$$

The relationship in the data booklet written as
$L = mrv = mr^2\omega$ refers to a **single particle only**.

Conservation of angular momentum

The **angular momentum** of a rotating object is **conserved** provided there are **no external torques** acting on the object.

Worked example

A turntable of moment of inertia $6.2 \times 10^{-3}\,\text{kgm}^2$ rotates freely with no friction at 5·2 rads⁻¹. A small mass of 0·22 kg is dropped onto the turntable at a distance of 0·13 m from the axis of rotation. The mass stays at the same position after falling onto the turntable. Calculate the new angular velocity of the turntable and mass.

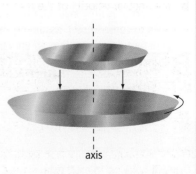

0·13m

$$I_{mass} = mr^2 = 0.22 \times 0.13^2 = 3.7 \times 10^{-3}\,\text{kgm}^2$$

Total angular momentum before = total angular momentum after

$$I_{turntable}\,\omega_1 = I_{turntable}\,\omega_2 + I_{mass}\,\omega_2$$
$$6.2 \times 10^{-3} \times 5.2 = (6.2 \times 10^{-3} + 3.7 \times 10^{-3})\,\omega_2$$
$$\omega_2 = 3.3\,\text{rads}^{-1}$$

DON'T FORGET

Remember to include $I\omega$ for the small mass.

Example 1

A disc of moment of inertia $7.5 \times 10^{-3}\,\text{kgm}^2$ rotates freely with no friction at 16 rads⁻¹. A second smaller disc of moment of inertia $4.2 \times 10^{-3}\,\text{kgm}^2$ drops onto the larger disc, with the centres of both discs on the axis of rotation.

axis

a Calculate the angular velocity of both discs after the impact, assuming both discs stick together and the smaller disc had no initial angular velocity.

b Show that rotational kinetic energy is not conserved, and calculate the amount of rotational kinetic energy converted into heat energy.

contd

ANGULAR MOMENTUM contd

Example 2

A horizontal disc rotates freely with negligible friction at 80 rpm. A small piece of putty of mass 2×10^{-2} kg falls vertically onto the disc and sticks to it at a distance of 0·18 m from the axis of rotation. The disc plus putty now rotates at 70 rpm. Calculate the moment of inertia of the disc.

Diver

Divers, ice skaters, acrobats and ballet dancers often make use of the conservation of angular momentum to increase or decrease their angular speed in a spin. A diver leaves the diving board with some initial angular velocity and with arms and legs outstretched. The axis of rotation is near the middle of his body. By pulling in her arms and legs, the diver's moment of inertia decreases as more of his mass is closer to the axis of rotation. Since I has gone down, then ω must increase to keep the angular momentum constant.

$$L_{start} = L_{later} \text{ (angular momentum conserved)}$$

$$I_{big} \, \omega_{small} = I_{smaller} \, \omega_{bigger} \text{ (with arms, legs pulled in)}$$

The diver now stretches out her arms and legs as she approaches the water. This increases I and reduces ω.

Note that, although the force of gravity acts on the diver, this force does not exert a torque on the diver. The requirement of no external torques is met.

Example 3

An ice skater rotates at 3 rads^{-1} with her arms outstretched. Each arm has a mass of 6 kg, and her total arm span is 1·8 m. By treating both arms as a single rod, show that the moment of inertia of her outstretched arms is 3·2 kgm^2. The moment of inertia of the rest of her body is 0·6 kgm^2.

The skater now wraps her arms around her body. The arms can now be treated like a solid disc of radius 0·30 m. Show that the new angular velocity of the skater is 10 rads^{-1}.

Example 4

A bullet of mass 10 g travelling at 80 ms^{-1} strikes the edge of a stationary disc tangentially and lodges in it. The disc has a mass of 0·75 kg and a radius of 0·30 m and is free to rotate about an axis through its centre. Calculate:

a the angular momentum of the bullet about the axis before impact.

b the angular velocity of the disc and bullet after impact.

Look up http://sparknotes.com/testprep/books/sat2/physics/chapter10section6. rhtml for another description of angular momentum.

LET'S THINK ABOUT THIS

Global warming will result in large sections of the polar ice caps melting. Explain what effects this will have on the rotation of the Earth, however slight.

GRAVITATION

INVERSE SQUARE LAW OF GRAVITATION

Newton's theory of gravitation states that the **gravitational force of attraction** between two objects is **directly proportional** to the **mass** of each object and is **inversely proportional** to the **square of their distance** apart.

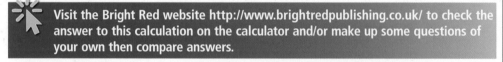

$F = \dfrac{Gm_1m_2}{r^2}$ where G is the **universal gravitational constant** with a value of

$$G = 6.67 \times 10^{-11} \, Nm^2kg^{-2}$$

> Enter "Cavendish-Boys" into an Internet search engine to find out how G was measured.

Astronomical data

Mass of Earth	6.0×10^{24} kg
Radius of Earth	6.4×10^6 m
Mass of Moon	7.3×10^{22} kg
Radius of Moon	1.7×10^6 m
Mean radius of Moon orbit	3.84×10^8 m
Mass of Jupiter	1.0×10^{27} kg
Radius of Jupiter	7.15×10^7 m
Mean radius of Jupiter orbit	7.8×10^{11} m

The gravitational force between the Earth and the Moon is:

$$F = \frac{Gm_1m_2}{r^2} = \frac{6.67 \times 10^{-11} \times 6 \times 10^{24} \times 7.3 \times 10^{22}}{3.84 \times 10^{8^2}} = 2.0 \times 10^{20} \, N$$

Example 1

The gravitational force between two people of masses 70 kg and 90 kg is 2.9×10^{-7} N. How far apart are the two people?

> Visit the Bright Red website http://www.brightredpublishing.co.uk/ to check the answer to this calculation on the calculator and/or make up some questions of your own then compare answers.

Example 2

Calculate the maximum and minimum gravitational force between the Earth and Jupiter as they rotate around the Sun.

GRAVITATIONAL FIELD STRENGTH

A **gravitational field** is a region where **gravitational forces** exist. **Gravitational field strength g** at a point is defined as the **force per unit mass** at that point. Or, more simply, it is the **force** on a **1 kg mass** placed at that point. In Higher Physics, you will have found that **g** on the surface of the Earth is **9.8 Nkg⁻¹**. This is consistent with the **inverse square law of gravitation** as shown:

Gravitational force per kg on Earth's surface

$$= \frac{Gm_1m_2}{r^2} = \frac{6.67 \times 10^{-11} \times 6 \times 10^{24} \times 1}{(6.4 \times 10^6)^2} = 9.8 \, N \text{ (per kilogram)}$$

contd

GRAVITATIONAL FIELD STRENGTH contd

g only has the value 9·8 Nkg⁻¹ at points on or near the Earth's surface. The value of *g* will **decrease** at a point high **above the Earth's surface** as *r* will be **greater**. The graph shows how *g* changes as *r* increases.

Example 3

Complete the table to show *g* decreasing with distance from centre of Earth.

$r / \times 10^6$ m	6·4	8	10	15	20
g/Nkg⁻¹	9·8				

Example 4

Calculate *g* on the surface of the Moon.

GRAVITATIONAL FIELD LINES AROUND MASSES

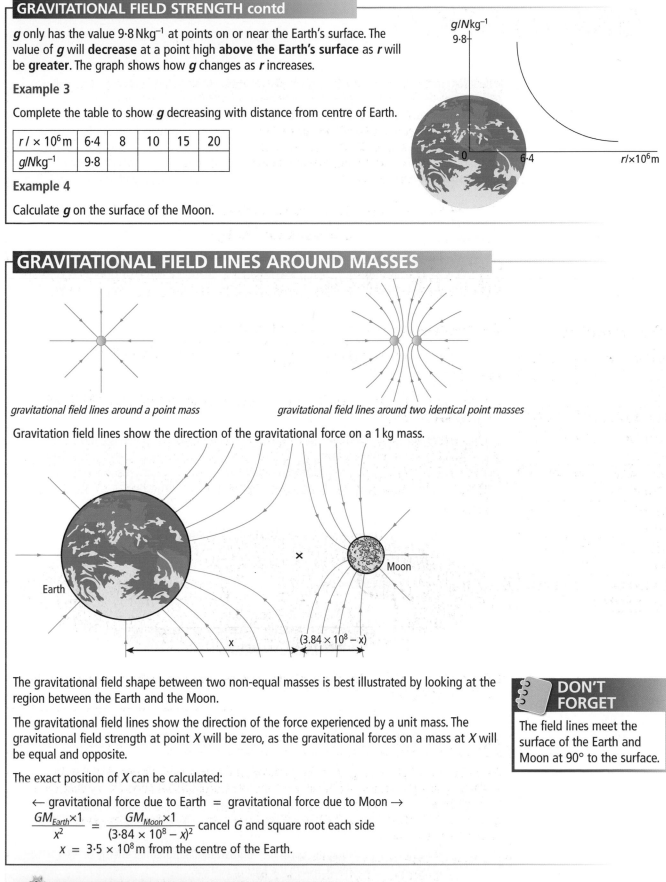

gravitational field lines around a point mass   *gravitational field lines around two identical point masses*

Gravitation field lines show the direction of the gravitational force on a 1 kg mass.

Earth   ×   Moon

x $(3.84 \times 10^8 - x)$

The gravitational field shape between two non-equal masses is best illustrated by looking at the region between the Earth and the Moon.

The gravitational field lines show the direction of the force experienced by a unit mass. The gravitational field strength at point *X* will be zero, as the gravitational forces on a mass at *X* will be equal and opposite.

The exact position of *X* can be calculated:

← gravitational force due to Earth = gravitational force due to Moon →

$$\frac{GM_{Earth} \times 1}{x^2} = \frac{GM_{Moon} \times 1}{(3.84 \times 10^8 - x)^2} \quad \text{cancel } G \text{ and square root each side}$$

$$x = 3.5 \times 10^8 \text{ m from the centre of the Earth.}$$

DON'T FORGET

The field lines meet the surface of the Earth and Moon at 90° to the surface.

LET'S THINK ABOUT THIS

1   A passenger of mass 70 kg is in an aeroplane flying at 13 000 m. Show that the force of gravity on the passenger is 3 N less at this height compared to the force of gravity at sea level.

2   Is the force of gravity on you from the Sun greater at midday or midnight? Explain your answer.

GRAVITATION

In earlier physics work, gravitational potential energy was calculated using the relationship $E_p = mgh$ where h is the vertical height above the Earth's surface. The potential energy at the Earth's surface was taken as zero, and the relationship is valid provided g remains constant. However, the force of gravity **does** change as we move away from the Earth, so a new expression for potential energy is required which can be applied at all points in space. To calculate **gravitational potential energy**, we need to introduce the new concept of **gravitational potential**, and use integration to calculate gravitational potential energy.

GRAVITATIONAL POTENTIAL

DON'T FORGET

The minus sign must be included.

Gravitational potential at a point in space is defined as the **work done** by external forces to move **unit mass** from **infinity** to that point. **Infinity** is where the **force of gravity** is **zero**. The **gravitational potential** V at a **distance** r from a **mass** m is given by:

$$V = -\frac{Gm}{r} \quad \text{the unit of } V \text{ is Jkg}^{-1}$$

GRAVITATIONAL POTENTIAL ENERGY

DON'T FORGET

This formula is **not** given in the data booklet.

The **gravitational potential energy** of **mass** m_1 at a distance r from another mass m is given by

$$E_p = -\frac{Gmm_1}{r} \quad \text{the unit of } E_P \text{ is J}$$

The minus sign appears during the integration and can be explained as follows: moving m_1 away from m requires an **increasing** amount of **work done**, overcoming the **attractive force** between the two masses. The **gravitational potential energy** must **increase to zero** at **infinity**, so it must be **negative** at **all points between m and infinity**.

Worked example

A satellite of mass 250 kg orbits the Earth at a height of 200 km. Calculate:

a the gravitational potential at this height

$$R_{orbit} = R_{Earth} + \text{height} = 6\cdot4 \times 10^6 + 200 \times 10^3 = 6\cdot6 \times 10^6\,\text{m}$$
$$V = -\frac{Gm}{r}$$
$$= -\frac{6\cdot67 \times 10^{-11} \times 6 \times 10^{24}}{6\cdot6 \times 10^6}$$
$$= -6\cdot06 \times 10^7\,\text{Jkg}^{-1}$$

DON'T FORGET

Minus signs must be carried forward at all stages.

b the gravitational potential energy of the satellite.

$$E_P = -\frac{Gmm_1}{r}$$
$$= -6\cdot06 \times 10^7 \times 250$$
$$= -1\cdot5 \times 10^{10}\,\text{J}$$

Example 1

Calculate the gravitational potential at a height of 150 km above the Moon's surface.

Example 2

A satellite of mass 750 kg orbiting the Earth has a gravitational potential energy of $-4\cdot4 \times 10^{10}$ J. Calculate the height of the satellite above the Earth's surface.

If a rocket is launched from Earth with sufficient kinetic energy, it will leave the Earth's gravitational field and not return. If the kinetic energy is not sufficient, then the rocket will fall back to Earth or stay in orbit round the Earth.

ESCAPE VELOCITY

The **escape velocity** from a planet is defined as the **minimum velocity** required to just **escape from the planet's gravitational field** and **reach infinity** with **zero velocity**.

The derivation of escape velocity of a mass m from a planet of radius r is as follows:

Total energy on planet's surface = total energy at infinity = 0

$$E_k + E_P = 0$$
$$\tfrac{1}{2}mv^2 + \left(-\frac{GMm}{r}\right) = 0 \quad \text{where } M \text{ is the mass of the planet}$$
$$v^2 = \frac{2GM}{r} \Rightarrow v = \sqrt{\frac{2GM}{r}}$$

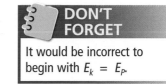

DON'T FORGET

It would be incorrect to begin with $E_k = E_P$.

Worked example

Calculate the theoretical escape velocity of a rocket fired from the surface of the Earth.
$$v = \sqrt{\frac{2GM}{r}} = \sqrt{\frac{2 \times 6\cdot67 \times 10^{-11} \times 6 \times 10^{24}}{6\cdot4 \times 10^6}} = 1\cdot1 \times 10^4\,\text{ms}^{-1}.$$

The effects of air resistance have not been included, and the rocket would need to reach this speed quite soon after launch. In practice, spacecraft deliberately escaping the Earth's gravitational field will be placed in orbit first (above the Earth's atmosphere) then given an escape velocity from the orbit (see Example 4).

Example 3 Calculate the escape velocity from the Moon.

Example 4 Calculate the escape velocity from an orbit 300 km above the Earth's surface.

Example 5

Calculate the escape velocity from a spherical asteroid of radius 150 km and density 2500 kgm^{-3}.

What happens when speeds are greater or less than the escape velocity is illustrated in the next two examples.

Worked example

Following a meteorite impact on the surface of the Moon, a rock of mass 12 kg is ejected vertically with a speed of $9\cdot5 \times 10^2$ ms^{-1}. Calculate the maximum height reached by the rock. Ignore the Earth's gravitational field.

Total energy of rock at start = $E_P + E_k$
$$E_P + E_k = -\frac{GMm}{r} + \tfrac{1}{2}m\,v^2 = -\frac{6\cdot67 \times 10^{-11} \times 7\cdot3 \times 10^{22} \times 12}{1\cdot7 \times 10^6} + 0\cdot5 \times 12 \times (9\cdot5 \times 10^2)^2$$
$$= -3\cdot44 \times 10^7 + 5\cdot4 \times 10^6 = -2\cdot9 \times 10^7\,\text{J}$$

When the rock stops moving, its total energy will be potential energy only, as the kinetic energy is zero.
$$-\frac{GMm}{r} = -2\cdot9 \times 10^7 \Rightarrow r = \frac{6\cdot67 \times 10^{-11} \times 7\cdot3 \times 10^{22} \times 12}{2\cdot9 \times 10^7} = 2 \times 10^6\,\text{m}$$
$$\text{height} = r - \text{Radius}_{\text{Moon}} = 2 \times 10^6 - 1\cdot7 \times 10^6 = 3\cdot0 \times 10^5\,\text{m} = 30\,\text{km}.$$

Example 6

If the rock in the above example had left the Moon's surface with a vertical speed of $2\cdot6 \times 10^3$ ms^{-1}, show that it will leave the Moon's gravitational field with a speed of $1\cdot02 \times 10^3$ ms^{-1}.

Black holes

A very dense star will have a massive gravitational field strength in the region around it. If the escape velocity is greater than 3×10^8 ms^{-1}, then photons of light will not be able to escape from this gravitational field. As we look in the direction of this star, no light from the star will be seen. The absence of any light from this direction gives rise to the term 'black hole'.

Photons of light from background stars passing close to this dense star can also be attracted by its massive gravitational field and move in curved paths rather than straight lines.

SATELLITE MOTION

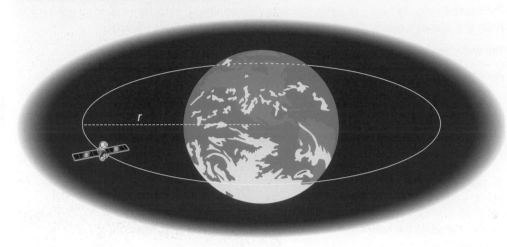

DON'T FORGET

A similar expression for ω^2 is another popular derivation in AH Physics.

Satellites are kept in orbit around a planet by the gravitational force, which provides the centripetal force on the satellite. Equating these two forces gives:

$$\frac{Gm_pm_s}{r^2} = m_s\omega^2r \qquad m_p = \text{mass of planet} \quad m_s = \text{mass of satellite}$$

$$\frac{Gm_pm_s}{r^2} = m_s 4\frac{\pi^2}{T^2}r \quad \text{substituting } \omega = \frac{2\pi}{T} \text{ then cancel } m_s$$

$$T^2 = \frac{4\pi^2}{Gm_p}r^3 \quad \text{Kepler's third law (not in the data booklet)}$$

Kepler's third law states that the **period** of a satellite **squared** is **directly proportional** to the **cube of the radius of the orbit**. This relationship is very useful for calculations involving satellite orbits.

Worked example

A satellite orbits 300 km above the Earth's surface. Calculate the time taken for one orbit of the Earth.

$$r = r_{\text{Earth}} + \text{height above Earth} = 6\cdot4 \times 10^6 + 300 \times 10^3 = 6\cdot7 \times 10^6 \text{m}$$

$$T^2 = \frac{4\pi^2}{Gm_p}r^3$$

$$= \frac{4 \times 3\cdot14^2 \times (6\cdot7 \times 10^6)^3}{6\cdot67 \times 10^{-11} \times 6 \times 10^{24}} = 2\cdot96 \times 10^7$$

$$\Rightarrow T = 5\cdot4 \times 10^3 \text{s} = 90.7 \text{ minutes.}$$

Example 1

A weather satellite orbits the Earth every 87 minutes. Calculate the height of the satellite above the Earth's surface.

Example 2

Calculate the height of a geostationary satellite above the Earth's equator.

Example 3

The Moon is a satellite of the Earth. Show that the Moon takes just under one month to orbit the Earth.

contd

SATELLITE MOTION contd

Example 4

Titan is the largest of Saturn's moons, with an orbit time of 15·9 days at a mean orbit radius of $1·22 \times 10^9$ m. Calculate the mass of Saturn.

Example 5

The International Space Station (ISS) makes around 15·7 orbits of the Earth per day. Calculate the height of the ISS above the Earth's surface, assuming circular orbits.

Example 6

Information about Jupiter's four Galilean moons is given in the table.

	Io	Europa	Ganymede	Calisto
orbit time T/days	1·77	3·55	7·16	16·7
mean orbit radius $R \times 10^8$ km	4·22	6·71	10·7	18·8

Show that T^2 is directly proportional to R^3.

Galileo saw these moons changing position over a period of time with his own basic telescope. You too can share the experience and see these Galilean moons with a pair of ordinary binoculars. To see when Jupiter is visible, and lots more, visit the Bright Red website http://www.brightredpublishing.co.uk/

Example 7

Calculate the linear speed of the Earth as it orbits the Sun.

Example 8

Show that the linear speed of an orbiting satellite is given by $v = \sqrt{\dfrac{GM}{R}}$ where M is the mass of the planet and R is the radius of the orbit.

✳ **Look up http://www.jodrellbank.manchester.ac.uk/astronomy/nightsky and http://ajdesigner.com/phpgravity/keplers_law_equation_period.php for information on Jupiter and to try more calculations on satellite orbits.**

⚙ LET'S THINK ABOUT THIS

The distance from the Earth to the Sun is often quoted as 93 million miles (1 mile = 1609 metres). Calculate the number of days in a year using these figures.

SIMPLE HARMONIC MOTION

In previous pages, linear and circular motion have been studied. The next type of motion we study is **oscillation** or **vibration**, where an object repeats a movement at **regular time intervals**. **Simple harmonic motion** (**SHM**) is a common type of oscillation where an object vibrates about an equilibrium position under the influence of an **unbalanced force** which is:

1 always directed **towards** the **equilibrium** position

2 **proportional** to the object's **displacement** from the equilibrium position.

Examples of SHM include:

guitar string

ends of a tuning fork

EQUATIONS FOR SHM

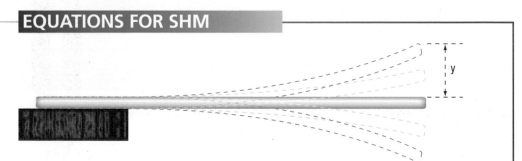

One end of a thin flexible rod vibrates vertically, performing SHM with **frequency** *f*. The **displacement** *y* of the end of the rod can be described by:

$$y = A\sin\omega t$$ where A is the amplitude of the vibration, $\omega = 2\pi f$ and $y = 0$ when $t = 0$

Velocity is defined as the **rate of change of displacement**.

The velocity of the rod end is given by:

$$v = \frac{dy}{dt}$$
$$= \frac{d}{dt}(A\sin\omega t)$$
$$= A\omega\cos\omega t$$

Acceleration is defined as the **rate of change of velocity**.

The acceleration of the rod end is given by:

$$a = \frac{dv}{dt}$$
$$= \frac{d}{dt}(A\omega\cos\omega t)$$
$$= -A\omega^2\sin\omega t$$
$$= -\omega^2 y$$

acceleration $a = -\omega^2 y$

DON'T FORGET

You could have started using $y = A\cos\omega t$ and repeated the steps to finish with $a = -\omega^2 y$.
$y = A\cos\omega t$ begins with the rod tip at $y = A$ when $t = 0$

---

contd

EQUATIONS FOR SHM contd

The maximum acceleration $= \pm \omega^2 A$ when $y = \pm A$

The velocity v of the rod end is:

$$v = A\omega\cos\omega t$$
$$= \pm A\omega \sqrt{1 - \sin^2\omega t} \qquad \text{since } \sin^2\omega t + \cos^2\omega t = 1$$
$$= \pm A\omega \sqrt{1 - \frac{y^2}{A^2}}$$
$$v = \pm \omega \sqrt{A^2 - y^2}$$

The maximum velocity equals $\pm\omega A$ and occurs when $y = 0$.

Worked example

The tip of one of the prongs of a tuning fork vibrates with SHM
of frequency 440 Hz and amplitude 0.6 mm.

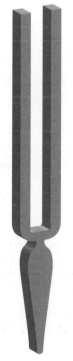

a Calculate the maximum acceleration and maximum speed of the tip.

First calculate ω.

$$\omega = 2\pi f$$
$$= 2 \times 3\cdot14 \times 440$$
$$= 2\cdot76 \times 10^3 \, rads^{-1}$$
$$a_{max} = \omega^2 A$$
$$= (2\cdot76 \times 10^3)^2 \times (0\cdot6 \times 10^{-3})$$
$$= 4\cdot6 \times 10^3 \, ms^{-2}$$
$$v_{max} = \omega A$$
$$= (2\cdot76 \times 10^3) \times (0\cdot6 \times 10^{-3})$$
$$= 1\cdot66 \, ms^{-1}$$

b Calculate the acceleration and speed of the tip when the displacement
of the tip is 0.25 mm from the equilibrium position.

$$a = -\omega^2 y$$
$$= -(2\cdot76 \times 10^3)^2 \times (0\cdot25 \times 10^{-3})$$
$$= -1\cdot9 \times 10^3 \, ms^{-2} \quad \text{(the tip is decelerating when } y = (+)0.25\,mm)$$
$$v = \pm \omega \sqrt{A^2 - y^2}$$
$$= 2\cdot76 \times 10^3 \times \sqrt{(0\cdot6 \times 10^{-3})^2 - (0\cdot25 \times 10^-)^2}$$
$$= 1\cdot5 \, ms^{-1}$$

Example 1

The displacement y of an object of mass 0.75 kg undergoing SHM is given by:

$$y = 0\cdot25\sin45t \text{ metres}$$

Calculate:

a the frequency of the oscillation

b the maximum unbalanced force applied to the object

c the speed of the object at a displacement of 0.18 m.

LET'S THINK ABOUT THIS

Not all repetitive periodic motion is SHM. The windscreen wipers on a car have a periodic motion
with a certain frequency but move with steady speed for most of the time.

The condition for SHM requires the acceleration of the blades to be proportional to the
displacement from a fixed point. The wiper blade motion is not SHM since the condition
$a = -\omega^2 y$ is not met.

OSCILLATING SPRING

A spring hanging vertically has a **mass m** attached to its lower end. This causes the spring to stretch and increase in length. The increase in length of the spring, or **extension**, is labelled e_1 on the diagram.

The spring **extension** is **directly proportional** to the **tension** in the spring.

The **tension** in the spring equals **mg** when the forces are balanced (spring and mass stationary).

$mg \propto$ extension

$mg = k \times$ extension, where **k** is the **constant of proportionality**

In the above example $k = \dfrac{mg}{e_1}$

Now pull the mass down, increasing the length by a second extension e_2.

Release the mass and it will perform SHM because the unbalanced force is proportional to the extension (displacement) of the mass from its original position.

$$\text{Unbalanced force} = -\frac{mg}{e_1} \times \text{extension}$$

$$ma = -\frac{mg}{e_1} \times \text{extension}$$

$$a = -\frac{g}{e_1} \times \text{extension}$$

The amplitude of the SHM will be e_2 and $\omega^2 = \dfrac{g}{e_1}$

Worked example

A mass of 0·6 kg causes an extension of 2 cm when hung on the end of a spring. The mass is pulled down a further 0·5 cm and released. Show that the frequency of oscillation is 3·5 Hz.

$$\omega^2 = \frac{g}{e_1}$$

$$= \frac{9 \cdot 8}{0 \cdot 02}$$

$$\omega = \sqrt{490}$$

$$= 22 \cdot 1 \text{ rads}^{-1}$$

$$f = \frac{\omega}{2\pi}$$

$$= \frac{22 \cdot 1}{2 \times 3 \cdot 14}$$

$$= 3 \cdot 5 \text{ Hz}$$

DON'T FORGET

Minus sign because the force and extension are in opposite directions.

SIMPLE PENDULUM

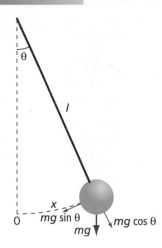

A popular class experiment in AH Physics is measuring *g* using a simple pendulum. The pendulum consists of a mass *m* on the end of a string of length *l* as shown.

The tangential component of the weight **mgsinθ** is the **unbalanced force** on the mass.

For small angles, $\sin\theta = \theta = \dfrac{arc\ length\ (x)}{l}$

The acceleration of the mass is:

$$a = \frac{unbalanced\ force}{m}$$
$$= -\frac{mg\frac{x}{l}}{m}$$
$$= -\frac{g}{l}x$$

The motion must be SHM, as the acceleration is proportional to the displacement and $\omega^2 = \dfrac{g}{l}$.

This leads to an expression for the **period *T*** of the pendulum:

$$T = \frac{2\pi}{\omega}$$
$$= 2\pi\sqrt{\frac{l}{g}} \quad \text{or } T^2 = \frac{4\pi^2 t}{g}$$

The period *T* is measured for several different lengths *l* of the pendulum, and a graph of *T²* against *l* is drawn. The graph should be a **straight line through the origin** with gradient $\dfrac{4\pi^2}{g}$.

    g can be calculated using $g = \dfrac{4\pi^2}{gradient}$

Example 1

Use the following data to draw an appropriate graph to find a value of *g*. Note that the time of ten complete swings has been given.

length l/m	0·50	0·70	0·90	1·2	1·5
10 periods (10T)/s	14·3	17·0	19·2	22·2	24·8

Example 2

Calculate the length of a pendulum whose period is 1 second. If you have time, try this experimentally and see if theory agrees with practice.

KINETIC ENERGY AND POTENTIAL ENERGY IN SHM

As the mass on the end of a spring oscillates up and down, there is a constant **interchange** of **kinetic energy** to **potential energy** and vice versa. When the mass momentarily stops at the top and bottom of the oscillation, all the energy is potential energy. As the mass passes through the equilibrium position at maximum speed, all its energy is kinetic energy.

If no energy is lost to frictional forces, then the **total energy** ($E_k + E_p$) will be **constant**.

Expression for kinetic energy

A mass **m** oscillates in a straight line about **0** with an amplitude **A**. The motion is SHM. An expression for the kinetic energy E_k of mass **m** is derived as follows:

$$E_k = \tfrac{1}{2}mv^2 = \tfrac{1}{2}m(\pm\omega\sqrt{A^2 - y^2})^2 = \tfrac{1}{2}m\omega^2(A^2 - y^2)$$
$$E_k = \tfrac{1}{2}m\omega^2(A^2 - y^2)$$

Note that when **y = 0**, the **kinetic energy** is **maximum**:

$$(E_k)_{max} = \tfrac{1}{2}m\omega^2 A^2$$

The total energy of the system E_{Total} will also equal $\tfrac{1}{2}m\omega^2 A^2$

When **y = A**, the **kinetic energy** is **zero**:

$$(E_k)_{min} = \tfrac{1}{2}m\omega^2(A^2 - y^2) = \tfrac{1}{2}m\omega^2(A^2 - A^2) = 0$$

Expression for potential energy

An expression for the **potential energy E_p** of a mass m undergoing SHM is derived as follows:

$$(E_p)_{max} = (E_k)_{max} = \tfrac{1}{2}m\omega^2 A^2 \quad \text{when } y = A.$$

At displacement **y**, $E_p = E_{Total} - E_k$

$$= \tfrac{1}{2}m\omega^2 A^2 - \tfrac{1}{2}m\omega^2(A^2 - y^2)$$
$$= \tfrac{1}{2}m\omega^2 A^2 - \tfrac{1}{2}m\omega^2 A^2 - (-\tfrac{1}{2}m\omega^2 y^2)$$
$$E_p = \tfrac{1}{2}m\omega^2 y^2$$

The following graphs show how the kinetic and potential energies vary with displacement and time for a mass undergoing SHM.

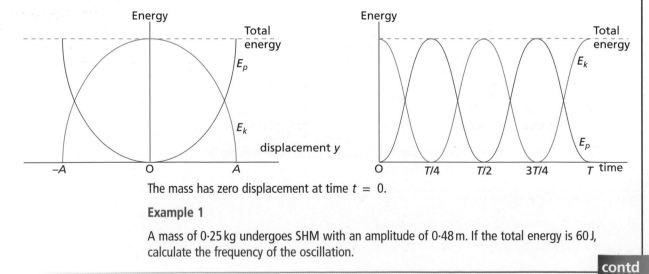

The mass has zero displacement at time $t = 0$.

Example 1

A mass of 0·25 kg undergoes SHM with an amplitude of 0·48 m. If the total energy is 60 J, calculate the frequency of the oscillation.

contd

KINETIC ENERGY AND POTENTIAL ENERGY IN SHM contd

Example 2

The displacement of a mass undergoing SHM is exactly half the amplitude at one instant. Show that the kinetic energy of the mass is 75% of the total energy at this instant.

Example 3

The graph shows how the potential energy of a 0·25 kg mass varies with displacement when performing SHM.

a What is the potential energy of the mass when the displacement is 0·2 m?

b What is the total energy of the mass?

c Calculate the kinetic energy of the mass when the displacement is 0·2 m.

d Calculate the speed of the mass when the displacement is 0·2 m.

e Calculate the maximum speed of the mass.

f What is the displacement when the speed is maximum?

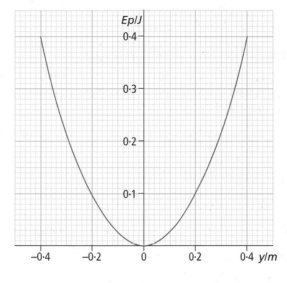

DAMPING

The amplitude of oscillation of a pendulum will gradually decrease to zero due to air resistance on the moving parts of the pendulum. This is called **damping**. The total energy of the pendulum will decrease to zero. This energy will be transformed into heat energy of the surroundings.

The following graphs show how the displacement varies with time for undamped oscillations and for lightly damped oscillations.

Notice that the amplitude decreases with each half damped oscillation.

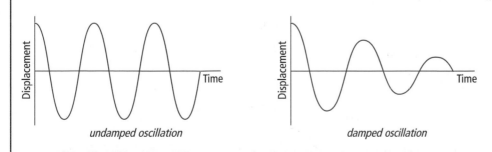

undamped oscillation   *damped oscillation*

> **DON'T FORGET**
>
> It would be **incorrect** to say the **speed** decreases as a result of damping.

LET'S THINK ABOUT THIS

Make use of the data book when using the formulae for F_μ and F_k.

Maximum E_p is when $y = A$, and maximum E_k is when $y = 0$.

The **total** energy is either E_p (max) or E_k (max).

WAVE–PARTICLE DUALITY

In the early 20th century, physicists had to come up with new theories to explain unexpected results found during some experiments. These new theories included:

- **electromagnetic radiation** having **particle** properties as well as **wave** properties

- **electrons** having **wave** properties as well as **particle** properties.

Electrons and **electromagnetic radiation** are said to exhibit **wave–particle duality**.

EVIDENCE OF WAVE–PARTICLE DUALITY

Electrons as particles

An **electron** has **mass** and can be **accelerated**. These properties suggest that electrons are **particles**.

Electron diffraction

Electrons can be shown to **diffract** and produce **interference patterns**. A diffraction tube like this has a thin graphite film in front of the electron gun, and electrons passing through will diffract and interfere in the region beyond the electron gun.

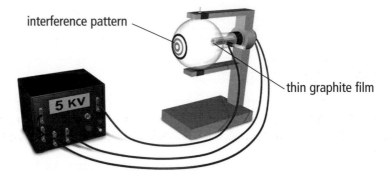

interference pattern

thin graphite film

5 KV

A series of concentric interference fringes will be seen on the fluorescent screen.

G.P. Thomson in Aberdeen was one of the first physicists to demonstrate electrons producing an interference pattern. An interference pattern is the test for a wave.

Interference patterns in electromagnetic radiation

Electromagnetic radiation can produce **interference patterns**; for example, visible light passing through a double slit produces a series of interference fringes, as was seen in Higher Physics. This provides evidence that electromagnetic radiation has wave properties.

The photoelectric effect

The **photoelectric effect** provides evidence of **electromagnetic radiation** having **particle** properties. You should be familiar with the details of the photoelectric effect from your studies in Higher Physics. Electromagnetic radiation falling on a metal surface can knock electrons out from the surface of the metal. The existence of a threshold frequency cannot be explained by wave theory alone. Einstein's solution to this anomaly was to suggest that light is made up of individual photons which have particle properties and whose energy is proportional to frequency. All observations of photoemission from the metal could now be explained.

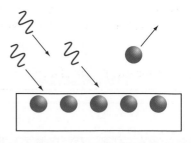

However, **photons** can still cause **interference patterns** and therefore have **wave properties**, hence the term **wave–particle duality**.

contd

WAVE–PARTICLE DUALITY

Enter "wave–particle duality" into an internet search engine for more information

de Broglie Wavelength

An expression for wavelength of a photon can be derived as follows:

The energy of a photon is:

$$E = hf$$

$$= h \times \frac{c}{\lambda}$$

$$mc^2 = h \times \frac{c}{\lambda} \quad \text{substituting } E = mc^2$$

$$\lambda = \frac{h}{mc}$$

$$\lambda = \frac{h}{p} \quad \text{where } p = \textbf{momentum (mass} \times \textbf{velocity)}$$

de Broglie proposed that this expression could be applied to both particles and waves.

Worked example 1

Calculate the wavelength of an electron travelling at $6{\cdot}5 \times 10^6\,ms^{-1}$.

$$\lambda = \frac{h}{p}$$

$$= \frac{h}{mv}$$

$$= \frac{6{\cdot}63 \times 10^{-34}}{9{\cdot}11 \times 10^{-31} \times 6{\cdot}5 \times 10^6}$$

$$= 1{\cdot}1 \times 10^{-10}\,m$$

A wavelength of $1{\cdot}1 \times 10^{-10}\,m$ is comparable to the spacing between atoms, and so diffraction and interference effects are possible, as this wavelength is of the same order as the gaps between atoms.

Worked example 2

Calculate the wavelength of a bullet of mass 10 g travelling at $120\,ms^{-1}$.

$$\lambda = \frac{h}{p}$$

$$= \frac{h}{mv}$$

$$= \frac{6{\cdot}63 \times 10^{-34}}{10 \times 10^{-3} \times 120}$$

$$= 5{\cdot}5 \times 10^{-34}\,m$$

A wavelength of $5{\cdot}5 \times 10^{-34}\,m$ is much smaller than any physical dimensions available, and so diffraction and interference effects associated with this bullet cannot be observed.

Example 1

Calculate the wavelength of an electron travelling at 5% of the speed of light.

LET'S THINK ABOUT THIS

Electron A has twice as much kinetic energy as electron B. What is the ratio of de Broglie wavelengths of electron A to electron B?

BOHR MODEL OF THE ATOM

The model for the hydrogen atom suggested by Niels Bohr in 1913 had a **positive nucleus** (a proton), with an **electron** orbiting only in certain **allowed orbits**.

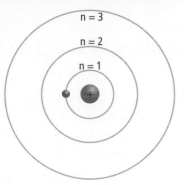

The **electron orbit circumference** had to contain a whole number (**n**) of de Broglie wavelengths.

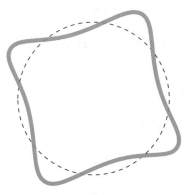

The fourth orbit would look like this, with four wavelengths fitting exactly into the orbit.

Bohr showed that the angular momentum of the orbiting electron is quantised in integer multiples of $\frac{h}{2\pi}$ as follows:

$$2\pi r = n\lambda$$
$$= n\frac{h}{p}$$
$$2\pi r = n\frac{h}{mv}$$
$$mvr = n\frac{h}{2\pi}$$

The angular momentum of the orbiting electron (**mvr**) is an integer multiple of $\frac{h}{2\pi}$.

Worked example 1

a Calculate the angular momentum of an electron in the second orbit of a hydrogen atom.

$$L = n\frac{h}{2\pi}$$
$$= 2 \times \frac{6.63 \times 10^{-34}}{2 \times 3.14}$$
$$= 2.1 \times 10^{-34}\,\text{kgm}^2\text{s}^{-1}$$

b If the speed of the electron is $1.1 \times 10^6\,\text{ms}^{-1}$ in the second orbit, calculate the radius of the second orbit.

$$mvr = n\frac{h}{2\pi} = 2.1 \times 10^{-34}$$
$$r = \frac{2.1 \times 10^{-34}}{9.11 \times 10^{-31} \times 1.1 \times 10^6}$$
$$= 2.1 \times 10^{-10}\,\text{m}$$

contd

BOHR MODEL OF THE ATOM contd

Example 1

Calculate the smallest angular momentum that an electron can have in a hydrogen atom.

Example 2

The angular momentum of an electron in the hydrogen atom is $4 \cdot 22 \times 10^{-34}\,\text{kgm}^2\text{s}^{-1}$. Which orbit is this electron in?

QUANTUM MECHANICS

Newtonian mechanics is perfectly satisfactory at predicting exactly what larger objects will do if we know their speed, position and forces acting on them.

This is not the case with atomic or sub-atomic particles, where measurements are difficult.

To find the exact position of an electron will require at least one photon of light interacting with the electron so that the observer can "see" it and perhaps carry out a measurement. But the photon reflecting off the electron will possibly cause the electron to change position or direction during the measurement.

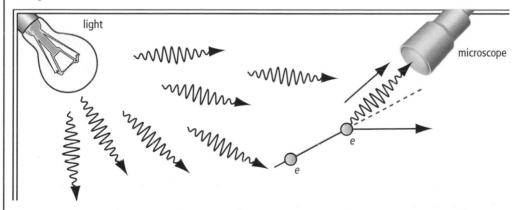

The diagram shows a photon of light reflecting off a moving electron into a microscope. The momentum of the photon causes the direction and velocity of the moving electron to change.

An exact prediction of what an atomic or sub-atomic particle will do is not possible. Instead, atomic and nuclear physicists treat particles like electrons as waves and can only predict what will happen in terms of probabilities and not certainties. Quantum mechanics provides methods to determine these probabilities.

Visit http://video.google.com/videoplay?docid=–5799457950699314965 for a video presentation on quantum mechanics.

LET'S THINK ABOUT THIS

The diagram on the opposite page showing four electron wavelengths fitting exactly into the fourth orbit circumference is relatively easy to draw.

Try drawing the third orbit with three wavelengths fitting exactly – still relatively easy to draw.

Now try drawing the second and first orbits with two and one wavelengths respectively fitting exactly on the circumferences – not so easy this time.

Visit www.upscale.utoronto.ca/PVB/Harrison/BohrModel/BohrModel.html

ELECTRIC FIELDS

COULOMB'S LAW

We have already found that two like charges exert a repulsive force on each other while two unlike charges exert an attractive force on each other.

Consider two point charges Q_1 and Q_2 separated by a distance r.

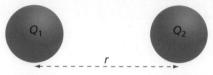

Coulomb found that the force F between the charges is **directly proportional** to the **magnitude** of each **charge** and **inversely proportional** to the **square** of the **distance** separating the charges.

$$F \propto \frac{Q_1 Q_2}{r^2}$$

$$F = \frac{1}{4\pi\varepsilon_0} \frac{Q_1 Q_2}{r^2} \qquad \text{where } \frac{1}{4\pi\varepsilon_0} \text{ is the } \textbf{constant of proportionality}.$$

The constant ε_0 is called the **permittivity of free space** and has a value of $8.85 \times 10^{-12} \, C^2 N^{-1} m^{-2}$.

The constant of proportionality $\frac{1}{4\pi\varepsilon_0}$ has a value $9 \times 10^9 \, Nm^2 C^{-2}$.

Worked example

The diagram shows two positive point charges separated by a distance of 0.15 m. Calculate the magnitude and direction of the electrostatic force exerted on the $+3.0\,\mu C$ charge.

$$F = \frac{1}{4\pi\varepsilon_0} \frac{Q_1 Q_2}{r^2}$$

$$= 9 \times 10^9 \frac{(3 \times 10^{-6}) \times (2 \times 10^{-6})}{0.15^2}$$

$$= 2.4 \, \text{N to the left as two positive charges repel.}$$

Note: the force on the $+2\,\mu C$ will be 2.4 N to the right.

Example 1

Calculate the magnitude and direction of the force on the smaller point charge in (a) and (b).

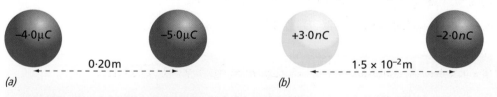

DON'T FORGET

The direction is found using 'like charges repel' and 'unlike charges attract'.

(a)   (b)

Electrostatic force between three collinear point charges

The **force on a point charge** due to the presence of two other point charges is the **vector sum** of the two forces caused by these other charges. The vector sum is straightforward if all three charges are collinear.

contd

COULOMB'S LAW contd

Worked example

Calculate the electrostatic force on the $-3\,\mu C$ point charge as shown in the diagram.

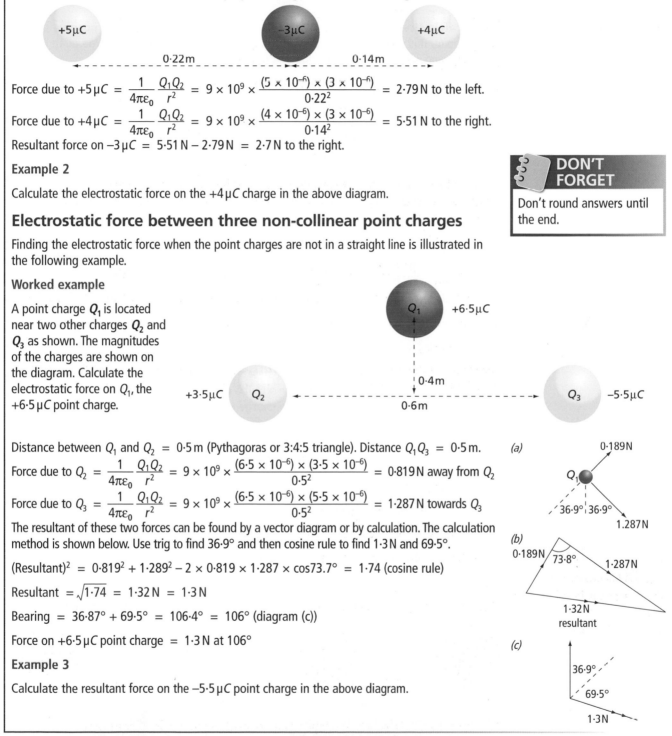

+5μC   –3μC   +4μC

0·22m   0·14m

Force due to $+5\,\mu C = \dfrac{1}{4\pi\varepsilon_0}\dfrac{Q_1Q_2}{r^2} = 9 \times 10^9 \times \dfrac{(5 \times 10^{-6}) \times (3 \times 10^{-6})}{0\cdot22^2} = 2\cdot79\,\text{N}$ to the left.

Force due to $+4\,\mu C = \dfrac{1}{4\pi\varepsilon_0}\dfrac{Q_1Q_2}{r^2} = 9 \times 10^9 \times \dfrac{(4 \times 10^{-6}) \times (3 \times 10^{-6})}{0\cdot14^2} = 5\cdot51\,\text{N}$ to the right.

Resultant force on $-3\,\mu C = 5\cdot51\,\text{N} - 2\cdot79\,\text{N} = 2\cdot7\,\text{N}$ to the right.

Example 2

Calculate the electrostatic force on the $+4\,\mu C$ charge in the above diagram.

Electrostatic force between three non-collinear point charges

Finding the electrostatic force when the point charges are not in a straight line is illustrated in the following example.

Worked example

A point charge Q_1 is located near two other charges Q_2 and Q_3 as shown. The magnitudes of the charges are shown on the diagram. Calculate the electrostatic force on Q_1, the $+6\cdot5\,\mu C$ point charge.

Q_1 +6·5μC

+3·5μC Q_2 0·4m Q_3 –5·5μC

0·6m

Distance between Q_1 and $Q_2 = 0\cdot5\,\text{m}$ (Pythagoras or 3:4:5 triangle). Distance $Q_1Q_3 = 0\cdot5\,\text{m}$.

Force due to $Q_2 = \dfrac{1}{4\pi\varepsilon_0}\dfrac{Q_1Q_2}{r^2} = 9 \times 10^9 \times \dfrac{(6\cdot5 \times 10^{-6}) \times (3\cdot5 \times 10^{-6})}{0\cdot5^2} = 0\cdot819\,\text{N}$ away from Q_2

Force due to $Q_3 = \dfrac{1}{4\pi\varepsilon_0}\dfrac{Q_1Q_2}{r^2} = 9 \times 10^9 \times \dfrac{(6\cdot5 \times 10^{-6}) \times (5\cdot5 \times 10^{-6})}{0\cdot5^2} = 1\cdot287\,\text{N}$ towards Q_3

The resultant of these two forces can be found by a vector diagram or by calculation. The calculation method is shown below. Use trig to find $36\cdot9°$ and then cosine rule to find $1\cdot3\,\text{N}$ and $69\cdot5°$.

$(\text{Resultant})^2 = 0\cdot819^2 + 1\cdot289^2 - 2 \times 0\cdot819 \times 1\cdot287 \times \cos73\cdot7° = 1\cdot74$ (cosine rule)

$\text{Resultant} = \sqrt{1\cdot74} = 1\cdot32\,\text{N} = 1\cdot3\,\text{N}$

$\text{Bearing} = 36\cdot87° + 69\cdot5° = 106\cdot4° = 106°$ (diagram (c))

Force on $+6\cdot5\,\mu C$ point charge $= 1\cdot3\,\text{N}$ at $106°$

Example 3

Calculate the resultant force on the $-5\cdot5\,\mu C$ point charge in the above diagram.

(a) 0·189N Q_1 36·9° 36·9° 1.287N

(b) 0·189N 73·8° 1·287N 1·32N resultant

(c) 36·9° 69·5° 1·3N

> **DON'T FORGET**
>
> Don't round answers until the end.

Visit http://hyperphysics.phy–astr.gsu.edu/HBASE/electric/elefor.html for an online Coulomb's Law calculator.

LET'S THINK ABOUT THIS

1 Use 9×10^9 for $\dfrac{1}{4\pi\varepsilon_0}$ rather than separate substitutions for π and ε_0.

2 Work out the **direction** of the force using your knowledge of attraction and repulsion.

ELECTRICAL PHENOMENA

ELECTRIC FIELD STRENGTH

As described earlier, a **field** in physics is a **region where forces exist**. Charged particles exert electrostatic forces on each other, so the **region around charged particles is called an electric field**.

DON'T FORGET

Electric field strength is analogous to gravitational field strength $(g = \frac{F}{m})$

Electric field strength E at a particular point is defined as the **force per unit positive charge** at that point. This means it is the **electrostatic force exerted on a point charge** of +1 C at that point and will require a **magnitude** and **direction**.

The magnitude of E can be calculated using the relationship:

$$E = \frac{F}{Q} \quad \text{The unit of } E \text{ is } NC^{-1}$$

Worked example

An electron moves between the cathode and anode in an electron gun. At one particular point, the electron experiences a force of $7 \cdot 2 \times 10^{-17}$ N towards the positive anode. Calculate the electric field strength at that point. The charge on an electron is $-1 \cdot 6 \times 10^{-19}$ C.

$$E = \frac{F}{Q}$$
$$= \frac{7 \cdot 2 \times 10^{-17}}{7 \cdot 2 \times 10^{-19}}$$
$$= 450 \, NC^{-1}$$

electron

cathode (–)   anode (+)

The direction of the electric field at this point will be towards the cathode, as a unit positive charge will experience a force towards the negative cathode.

Electric field around a point charge

+Q ◄ - - - - - - - - ► X

r

Point X is a distance r from a **point charge $+Q$**. The **electric field strength** at point X will be the **force experienced by a unit positive charge** (+1 C) placed at X. Coulomb's Law is used to calculate the force between $+Q$ and **+1 C** a distance r apart:

$$F = \frac{1}{4\pi\varepsilon_0} \frac{Q_1 Q_2}{r^2}$$
$$= \frac{1}{4\pi\varepsilon_0} \frac{Q \times 1}{r^2}$$
$$E = \frac{Q}{4\pi\varepsilon_0 r^2} \quad \text{since } E \text{ is the force on a unit positive charge.}$$

DON'T FORGET

Don't confuse nC and μC.

Example 1

Calculate the electric field strength at (a) 30 cm and (b) 40 cm from a point charge of $5 \cdot 0\,\mu C$.

Example 2

At what distance from a $6 \cdot 0\,nC$ point charge will the electric field strength be $110\,NC^{-1}$?

Electric field lines

A diagram of the electric field lines around a positive point charge is shown here:

The **radial field decreases** in value as **r increases**. A **negative point charge** will have the field lines pointing **towards the negative charge**.

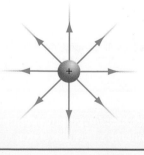

DON'T FORGET

The field lines must start on the charge surface and at right angles to the surface. Two field lines starting at the same point on the surface is incorrect.

contd

Visit http://www.ibiblio.org/links/devmodules/ElectricField/index.html for an interactive exercise on electric field patterns.

ELECTRIC FIELD STRENGTH contd

Electric field strength between two point charges

The electric field between two point charges depends on the polarity of the charges, as shown.

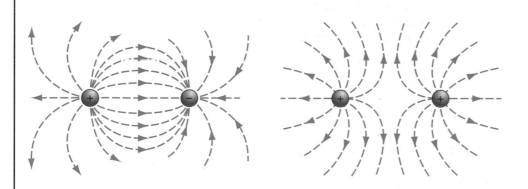

The arrows show the resultant force direction on a unit positive charge. The electric field strength at a particular point between the charges has a contribution from each charge. The following example shows how to calculate each contribution.

Worked example

Two point charges are separated by a distance of **6 cm**. Calculate the electric field strength at point **X**.

$$E \text{ at } X \text{ due to } -3.0\,\mu C = \frac{1}{4\pi\varepsilon_0}\frac{Q}{r^2} = 9 \times 10^9 \times \frac{(3 \times 10^{-6})}{0.04^2} = 1.69 \times 10^7\,NC^{-1} \text{ to the left.}$$

$$E \text{ at } X \text{ due to } +5.0\,\mu C = \frac{1}{4\pi\varepsilon_0}\frac{Q}{r^2} = 9 \times 10^9 \times \frac{(5 \times 10^{-6})}{0.02^2} = 1.125 \times 10^8\,NC^{-1} \text{ to the left.}$$

$$E \text{ at point } X = 1.69 \times 10^7 + 1.125 \times 10^8 = 1.3 \times 10^8\,NC^{-1} \text{ to the left.}$$

Example 3

Point Y is 3 cm beyond the $+5\,\mu C$ charge. Calculate the electric field strength at point Y.

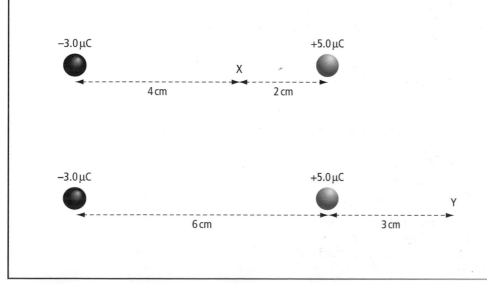

ELECTRICAL PHENOMENA

CHARGING BY INDUCTION

A metal sphere on an insulated base can be charged positively or negatively by a four-stage process known as "charging by induction". The sphere can be charged positively as shown:

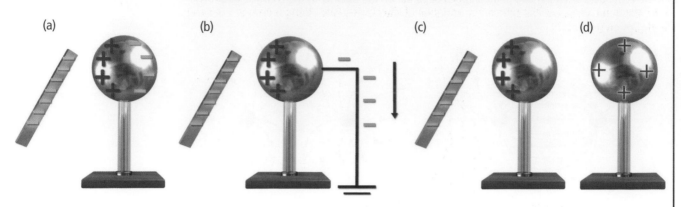

(a)   (b)   (c)   (d)

(a) Bring a negatively charged rod close to the sphere. Note the negative charges (electrons) on the surface of the sphere are repelled to the furthest side of the sphere with a net positive charge closest to the rod.

(b) Touch the sphere with a finger (to create an earth), keeping the charged rod close to the sphere. Some of the electrons on the sphere will flow to earth through your finger.

(c) Remove the earth (take your finger off the sphere), keeping the negative rod in position. This prevents the electrons returning to the sphere from earth.

(d) Now remove the charged rod. The sphere now has a net positive charge around its surface.

To charge the sphere negatively by induction requires a positively charged rod, and the four steps are repeated. Only the negative charges (electrons) can move, so step (b) will have the electrons moving from earth to the sphere.

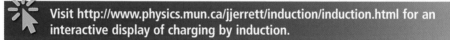

Visit http://www.physics.mun.ca/jjerrett/induction/induction.html for an interactive display of charging by induction.

Alternative method

An alternative three-stage method of charging 2 metal spheres is shown:

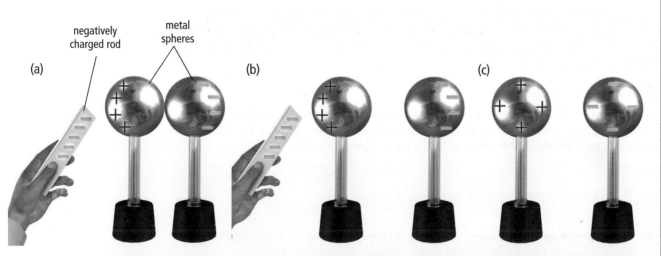

negatively charged rod   metal spheres

(a)   (b)   (c)

(a) Bring a negatively charged rod close to two touching isolated conducting spheres.

(b) Separate the spheres with the charged rod still close to the spheres. One sphere is charged positively (nearest to the negatively charged rod). The other sphere is charged negatively.

(c) Remove the charged rod, and the charges spread evenly over the surface of each sphere.

ELECTRIC FIELD AROUND A CHARGED CONDUCTING SPHERE

A conducting sphere of radius **a** is charged positively. The charge +*Q* will be evenly distributed over the outer surface of the sphere. The variation of the electric field strength around the sphere is shown on the graph.

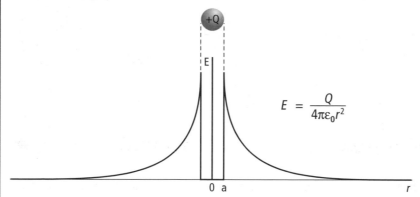

$$E = \frac{Q}{4\pi\varepsilon_0 r^2}$$

Outside the sphere, the electric field will follow an **inverse square relationship**. **Inside the sphere**, the value of *E* will be zero. No field lines exist inside the conducting sphere.

Conducting sphere in an electric field

A uniform electric field will be modified by placing a conducting sphere in it.

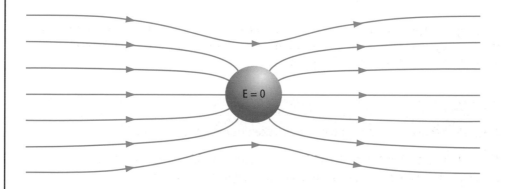

The electric field will induce charges on the surface of the conducting sphere. The field lines move from positive to negative and are perpendicular when they meet the sphere surface. Again, the **electric field will be zero inside the sphere**. If the sphere is hollow, the electric field is still zero inside the sphere.

Electrostatic shielding

Sensitive electronic devices are shielded from background electromagnetic interference by surrounding them with a conducting material. External electric fields have no influence inside the enclosure. Similarly, coaxial cable shields the inner cable with a metal mesh.

⚙ LET'S THINK ABOUT THIS

1   The inside of a car receives partial shielding from external electric fields. Mobile phone signals can enter through the window space.

2   The metal bodywork of a microwave oven prevents the electric field **inside** the oven from having any effect outside the oven. A honeycomb of conducting material is incorporated in the glass of the oven door.

ELECTROSTATIC POTENTIAL

Electrostatic potential is analogous to gravitational potential and is another way of giving information about an electric field.

The **electrostatic potential at a point** in an electric field is defined as the **work done** bringing a **unit positive charge** from **infinity** to **that point**.

Integration is used to derive the relationship for the **electrostatic potential V** at a distance r from a point charge $+Q$.

$$V = \frac{Q}{4\pi\varepsilon_0 r}$$ The unit of V is **volt V** or JC^{-1}

The **electrostatic potential at infinity is zero**, as is the case with gravitational potential, but there is no negative sign in this case. This is because a repulsive force has to be overcome as a unit positive charge is moved towards $+Q$, and work must be done.

A negative point charge of $-Q$ will have negative values of V around it. Work must be done as a unit positive charge moves away from $-Q$ to eventually reach zero volts at infinity.

Worked example

Point X is $0{\cdot}75$ m from a point charge of $+5\,\mu C$. Calculate the electrostatic potential at point X.

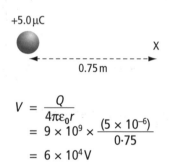

+5.0 μC

X

0.75 m

$$V = \frac{Q}{4\pi\varepsilon_0 r}$$
$$= 9 \times 10^9 \times \frac{(5 \times 10^{-6})}{0{\cdot}75}$$
$$= 6 \times 10^4\,V$$

Example 1

Calculate the electrostatic potential at a distance of 2 nm from an electron.

Electrostatic potential involving two or more point charges

When two or more point charges are involved, care must be taken to insert the minus sign for negative charges and to substitute the correct distance when calculating V.

Worked example

Calculate the electrostatic potential at point X between these two point charges.

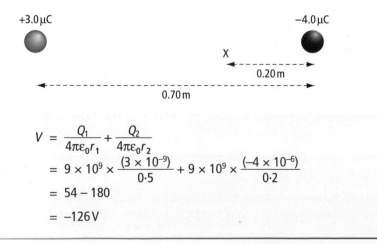

+3.0 μC   −4.0 μC

X

0.20 m

0.70 m

$$V = \frac{Q_1}{4\pi\varepsilon_0 r_1} + \frac{Q_2}{4\pi\varepsilon_0 r_2}$$
$$= 9 \times 10^9 \times \frac{(3 \times 10^{-9})}{0{\cdot}5} + 9 \times 10^9 \times \frac{(-4 \times 10^{-6})}{0{\cdot}2}$$
$$= 54 - 180$$
$$= -126\,V$$

contd

ELECTROSTATIC POTENTIAL contd

Example 2

Calculate the electrostatic potential at points X and Y near these point charges.

Potential of a charged conducting sphere

A charge $+Q$ on an isolated conducting sphere is **uniformly distributed over its surface** due to the repulsion of like charges. These charges on the sphere's surface do not move, so all points on the surface and inside the sphere must be at the same potential.

The potential outside the sphere is the same as if the whole charge is concentrated at the centre. This was the case with gravitational potential and gravitational field strength around a planet when the whole mass was considered to be at the centre.

The variation of V around a sphere of radius a with charge $+Q$ is shown in this graph.

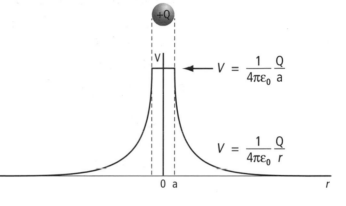

Note that the potential everywhere inside the sphere is $V = \dfrac{Q}{4\pi\varepsilon_0 a}$ but the electric field strength inside the sphere is zero.

Electric potential energy

A charged particle at a point in an electric field will have a potential energy E_p as work must be done **moving the charged particle from infinity to that point**. The electric potential energy of a charged particle can be calculated using the relationship:

$E_p = VQ$ where V is the potential at the point (V) and
                         Q is the charge (C)

Worked example

An alpha particle is $2.5\,\mu m$ from a point charge of $5.5\,nC$. Calculate the potential energy of the alpha particle at this position. The charge on an alpha particle is $3.2 \times 10^{-19}\,C$

$$E_p = VQ$$
$$= \frac{Q}{4\pi\varepsilon_0 r} \times Q_{alpha}$$
$$= 9 \times 10^9 \times \frac{(5\cdot5 \times 10^{-9})}{(2\cdot5 \times 10^{-6})} \times 3\cdot2 \times 10^{-19}$$
$$= 6\cdot3 \times 10^{-12}\,J$$

Example 3

Calculate the electric potential energy of a proton at a distance of $3\cdot8 \times 10^{-10}$ m from an alpha particle.

LET'S THINK ABOUT THIS

1   A negative charge should have the minus sign included in the calculation for electrostatic potential.

2   Electrostatic potential does not require a direction.

ELECTRICAL PHENOMENA

A **uniform field** exists between **two charged parallel plates** as shown.

An expression for the electric field strength between the plates can be derived as follows:

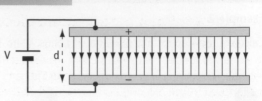

DON'T FORGET

The unit of E can be Vm^{-1} or NC^{-1}.

Two parallel plates have a **potential difference V** across them and are a **distance d** apart as shown. The electric field lines have a **direction from + to −**. Work must be done moving a charge $+Q$ from the bottom plate to the top plate, as a **repulsive force F** must be overcome:

$$\text{Work done} = F \times d$$

An alternative expression is:

$$\text{Work done} = QV$$

equating gives $\quad F \times d = QV$

rearranging $\quad \dfrac{F}{Q} = \dfrac{V}{d}$

substituting $\quad E = \dfrac{F}{Q}$ gives $E = \dfrac{V}{d}$

Example 1

The electric field strength between two parallel plates is $2 \cdot 5 \times 10^4 \, NC^{-1}$ when the *p.d.* across the plates is 2 kV. How far apart are the plates?

Worked example

Two parallel plates are 25 cm apart and have a potential difference of 2 kV applied across them as shown.

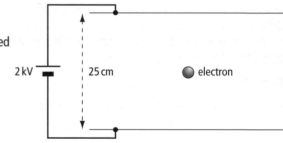

a Calculate the unbalanced force on an electron midway between the plates.

First calculate E: $\quad E = \dfrac{V}{d} = \dfrac{2000}{0.25} = 8 \times 10^3 \, Vm^{-1}$

Force due to electric field: $\quad F_{electric} = EQ = (8 \times 10^3) \times (1 \cdot 6 \times 10^{-19}) = 1 \cdot 3 \times 10^{-15} \, N$ (upwards)

The weight of the electron $= mg = 9 \cdot 11 \times 10^{-31} \times 9 \cdot 8 = 8 \cdot 9 \times 10^{-30} \, N$ (downwards)

The weight of the electron is very much less than $F_{electric}$, so the weight can be ignored.

$$\text{unbalanced force} = 1 \cdot 3 \times 10^{-15} \, N \text{ (upwards)}$$

b The parallel plates are in a vacuum, and the electron would accelerate upwards, as there are no collisions with air molecules. Calculate:

i the electron's acceleration.

acceleration: $\quad a = \dfrac{F}{m} = \dfrac{1 \cdot 3 \times 10^{-15}}{9 \cdot 11 \times 10^{-31}}$

$$= 1 \cdot 4 \times 10^{15} \, ms^{-2} \text{ (upwards)}$$

ii the time taken to reach the top plate.

time to move 12·5 cm from rest: $\quad s = ut + \dfrac{1}{2}at^2$

$$0 \cdot 125 = 0 + 0 \cdot 5 \times 1 \cdot 4 \times 10^{15} \times t^2$$

$$t = 1 \cdot 3 \times 10^{-8} \, s$$

contd

UNIFORM ELECTRIC FIELD contd

Two-dimensional motion in an electric field

The previous example considered a stationary electron in a uniform electric field. Now we will consider another common situation where a charged particle has an initial speed before entering the electric field. The region between the plates is evacuated.

charged particle

There is no **horizontal** force on this charged particle, so it will continue to move horizontally with a steady speed (Newton's First Law).

An **electric force** will act **vertically** on the particle. If the **charge on the particle is positive**, it will experience an electric force **towards the negative plate** and will **accelerate downwards** (Newton's Second Law). The resultant trajectory of a positively charged particle is shown by the red dotted line.

A **negatively charged particle** projected between the plates would **accelerate upwards** as well as continuing with a steady speed horizontally and would follow a path similar to the blue dotted line.

The exact trajectory depends on the mass, charge and speed of the particle as well as the separation and voltage of the plates.

evacuated tube

Worked example

Electrons enter an electric field midway between two deflecting plates with a speed of $2.1 \times 10^7 \, ms^{-1}$. The plates are 80 mm long and 50 mm apart and there is a potential difference of 500 V across the plates.

Calculate:

a the time an electron takes to pass the deflecting plates.

$$t = \frac{d}{v} = \frac{80 \times 10^{-8}}{2.1 \times 10^7} = 3.8 \times 10^{-9} \, s$$

b the vertical deflection **s** of an electron as it leaves the space between the deflecting plates.

vertical force on electron: $\quad F = EQ = \frac{v}{d} \times Q = \frac{500 \times 1.6 \times 10^{-19}}{50 \times 10^{-3}} = 1.6 \times 10^{-15} \, N$

vertical acceleration of electron: $\quad a = \frac{F}{m} = \frac{1.6 \times 10^{-15}}{9.11 \times 10^{-31}} = 1.756 \times 10^{15} \, ms^{-2}$

vertical displacement: $\quad s = ut + \frac{1}{2}at^2$

$$= 0 + 0.5 \times (1.756 \times 10^{15}) \times (3.8 \times 10^{-9})^2$$

$$= 1.27 \times 10^{-2} \, m$$

$$= 1.3 \, cm$$

Example 2

Calculate **s** in the above diagram if the electron speed is changed to $9.5 \times 10^6 \, ms^{-1}$ and the plate voltage is increased to 750 V.

> **DON'T FORGET**
>
> An electron has steady speed horizontally.

LET'S THINK ABOUT THIS

1 Charged particles move with steady speed parallel to the plates. Use $v = \frac{d}{t}$.

2 Charged particles accelerate uniformly towards the plate of opposite polarity. Use equations of motion as well as knowledge of the electric field strength between the plates.

3 If the speed of the charged particle is greater than 10% of the speed of light, then relativistic effects must be taken into account when doing calculations.

ELECTRICAL PHENOMENA

RUTHERFORD SCATTERING

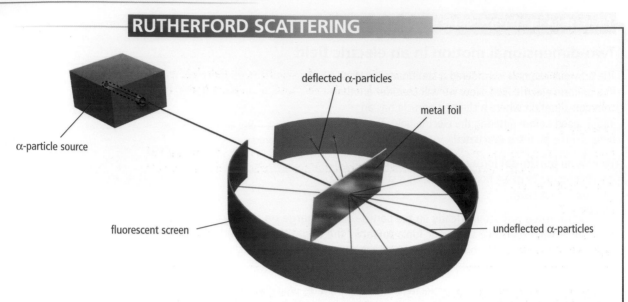

α-particle source

deflected α-particles

metal foil

fluorescent screen

undeflected α-particles

In 1909, Geiger and Marsden fired a beam of alpha particles at a thin gold foil – and, although most passed straight through the gold foil, some were deflected to the side. A few were even deflected backwards.

These results confirmed the nuclear model of the atom, with a positive nucleus exerting a repulsive force on any positive alpha particles which came close to the nucleus.

A very small number of alpha particles would be repelled straight back in the direction they had come from. One of these alpha particles must have approached a gold nucleus head-on, and the repulsive force would have slowed it down until it stopped momentarily before speeding up back along its original path. The kinetic energy of this alpha particle had been turned into electric potential energy when it stopped. The potential energy then turned back into kinetic energy as the alpha particle was repelled backwards.

The physics of a head-on "collision" of an alpha particle with a gold nucleus will now be studied in detail.

Worked example

An alpha particle approaches a gold nucleus with a speed of $8 \cdot 5 \times 10^5 \, \text{ms}^{-1}$. Calculate the distance of closest approach d for a head-on collision between the alpha particle and the gold nucleus.

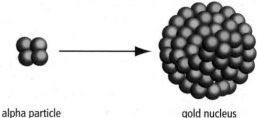

alpha particle

gold nucleus

The alpha particle has a mass of $6 \cdot 7 \times 10^{-27} \, \text{kg}$ and a charge of $2e = 2 \times 1 \cdot 6 \times 10^{-19} \, \text{C}$.

Gold has atomic number 79, so the charge on the gold nucleus is $79e = 79 \times 1 \cdot 6 \times 10^{-19} \, \text{C}$.

The kinetic energy of the alpha particle is:

$$E_K = \frac{1}{2}mv^2$$

$$= 0 \cdot 5 \times (6 \cdot 7 \times 10^{-27}) \times (8 \cdot 5 \times 10^5)^2 = 2 \cdot 42 \times 10^{-15} \, \text{J}.$$

The **total** energy of the alpha particle will be $2 \cdot 42 \times 10^{-15} \, \text{J}$ assuming $E_P = 0$ well away from the gold nucleus. When the alpha particle reaches its closest distance to the gold nucleus and stops, all the energy will be electric potential energy.

$$E_P = \frac{Q_{alpha} \times Q_{gold}}{4\pi\varepsilon_0 \times d}$$

$$2 \cdot 42 \times 10^{-15} = 9 \times 10^9 \times \frac{(3 \cdot 2 \times 10^{-19}) \times (79 \times 1 \cdot 6 \times 10^{-19})}{d}$$

$$d = 1 \cdot 5 \times 10^{-11} \, \text{m}.$$

contd

RUTHERFORD SCATTERING contd

Example 1

An α particle approaches a tungsten nucleus with a speed of $2.4 \times 10^6\,ms^{-1}$. Calculate its distance of closest approach to the tungsten nucleus. (You will need information from a Periodic Table.)

Example 2

An α particle approaches an osmium nucleus and comes to rest momentarily at a distance of $4.5 \times 10^{-11}\,m$ from the nucleus. Calculate the initial speed of the α particle well away from the influence of the nucleus.

Example 3

An α particle approaches a nucleus with a speed of $7.6 \times 10^5\,ms^{-1}$. Its distance of closest approach to the nucleus is $1.87 \times 10^{-11}\,m$. Identify the nucleus.

MILLIKAN'S OIL-DROP EXPERIMENT

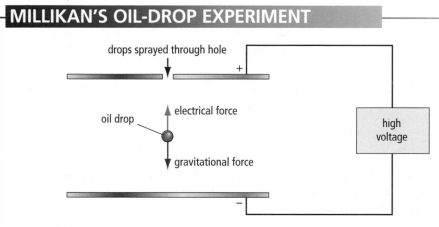

In 1909, Millikan sprayed charged oil droplets into an electric field between two parallel plates. By varying the voltage **V** across the plates, he was able to stop a particular droplet from moving downwards by balancing the weight of the oil drop **mg** with an upwards electric force **EQ** or $\frac{VQ}{d}$ where **Q** is the charge on the oil droplet and **d** is the separation of the plates:

$$mg = \frac{VQ}{d}$$

The mass of the oil drop was found after measuring its terminal velocity after **V** was switched off. The charge on the oil drop could be calculated when all the other variables were known. Millikan measured the charges on thousands of oil drops and noticed a pattern in the values of **Q**. He concluded that the values of electric charge were all integral multiples of **$1.6 \times 10^{-19}\,C$** (**e**). For example, $8.0 \times 10^{-19}\,C = 5e$ and $1.92 \times 10^{-18}\,C = 12e$. A charge of $4.0 \times 10^{-19}\,C$ is not possible as this would be $2\frac{1}{2}e$ and is not an integer multiple of **e**.

The term **quantum** is used to describe a quantity which exists in **integral multiples**, so we can say that **charge is quantised in units of $1.6 \times 10^{-19}\,C$**.

Visit http://physics.wku.edu/~womble/phys260/millikan.html for an interactive simulation of Millikan's experiment.

LET'S THINK ABOUT THIS

The formula for electric potential energy is not given in the data booklet. You must use:

$$E_p = Vq$$

or
$$E_p = \frac{Q_1 \times Q_2}{4\pi\varepsilon_0 r}$$
Remember that the denominator uses **r**, not r^2.

ELECTROMAGNETISM

MAGNETIC FIELDS

We are familiar with the concept of a magnetic field from earlier studies of physics, and we will now look in some detail at the **magnetic forces that arise on conductors** and **moving charged particles in magnetic fields**. The magnetic field lines show the direction that a compass needle would be forced to align with and **not** the direction of the force on any charges.

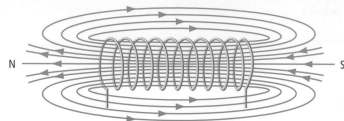

current-carrying coil

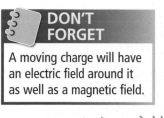

DON'T FORGET

A moving charge will have an electric field around it as well as a magnetic field.

The magnetic field lines travel from **north to south** by convention. When the current in the coil is switched off, the magnetic field disappears, so the **magnetic field** must be caused by the **moving charges** (electrons) in the wire.

The Earth's magnetic field is thought to be caused by electrons moving in its molten iron core by convection currents. The Earth's magnetic field is similar to a giant bar magnet.

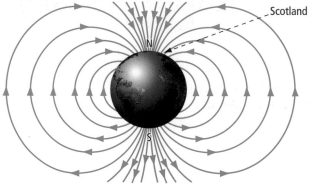

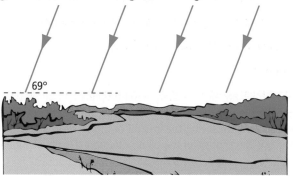

Looking east in Scotland

The Earth's magnetic field lines are horizontal at the Equator and vertical at the North Pole. In Scotland, the magnetic field lines are at an angle of approximately 69° to the horizontal. This angle increases with latitude.

Magnetic induction

The strength of a magnetic field at a point is called the **magnetic induction** and has the symbol **B**. The **unit of magnetic induction** is the **tesla (T)**. A more precise definition of magnetic induction **B** will follow.

FORCE ON A CURRENT-CARRYING CONDUCTOR

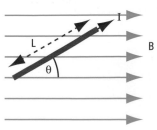

A conductor of length **L** carries a current **I** at an angle θ to a magnetic field of magnetic induction **B**. The conductor will experience a force given by the following expression:

$F = BIL\sin\theta$ (the derivation is not required)

When the conductor is perpendicular to the magnetic field, the force is:

$F = \boldsymbol{BIL}$ since θ = 90°

When the conductor is parallel to the magnetic field, the force is zero since θ = 0°.

contd

FORCE ON A CURRENT-CARRYING CONDUCTOR contd

Worked example

A conductor of length 60 cm carries a current in a magnetic field as shown. Calculate the magnitude of the force exerted on the conductor.

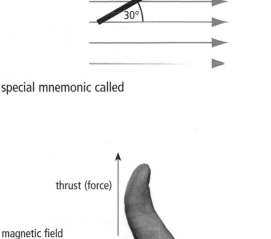

$$F = BIL\sin\theta$$

$$= (4 \cdot 5 \times 10^{-3}) \times 2 \cdot 5 \times 0 \cdot 6 \times \sin 30°$$

$$= 3 \cdot 4 \times 10^{-3} \, N$$

The direction of this force is **not** in the direction of **B** or the **current**. A special mnemonic called the **right-hand rule** is used to help predict the direction of this force.

Right-hand rule

The direction of the force on a current-carrying conductor in a magnetic field is:

- perpendicular to the plane carrying the conductor

- perpendicular to the direction of the magnetic field.

The right-hand rule uses the thumb and first two fingers of the right hand.

Arrange the thumb and first two fingers at right angles to each other as shown.

The **fi**rst finger represents the magnetic **fi**eld direction.

The **se**cond finger represents the **e**lectron current direction.

The **th**umb represents the **th**rust (force).

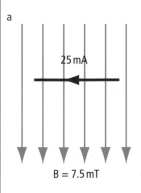

Example 1

A straight conductor of length 40 cm is placed in two different magnetic fields carrying different currents in each situation as shown. Calculate the magnitude and direction of the force on the conductor in each case.

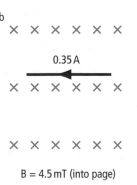

a
25 mA

B = 7.5 mT

b
× × × × × ×
0.35 A
× × × × × ×

× × × × × ×

B = 4.5 mT (into page)

Definition of magnetic induction

A magnetic field has a **magnetic induction** of **one tesla** when a conductor of **length one metre**, carrying a **current of one ampere** perpendicular to the field, is acted on by a **force of one newton**.

> **DON'T FORGET**
>
> $B = \dfrac{F}{IL\sin\theta}$: make everything equal to 1 for the definition. ($\theta = 90°$ for $\sin\theta = 1$)

LET'S THINK ABOUT THIS

Some physics textbooks will use "conventional current" where current direction is taken from positive to negative. These textbooks will use a different rule for the direction of the force on the conductor but will still give the same result as the right-hand rule.

ELECTRICAL PHENOMENA

ELECTRIC MOTOR

A simple electric motor consists of a single-turn coil of wire which can rotate about an axis in a magnetic field.

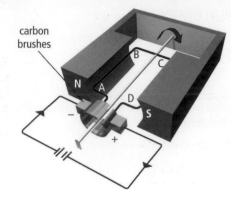

The coil is initially horizontal. When a current flows through the brushes into the coil, length **AB** experiences a force **F upwards**. (Use the right-hand rule to confirm the direction.) Length **CD** experiences a force **F downwards** as the electron current flows from C to D. Side **BC** experiences **no force** as the current is parallel to the magnetic field.

The coil experiences **two torques** which cause the coil to turn clockwise, as seen from the front.

$$\text{Torque} = F \times (\tfrac{1}{2}BC) + F \times (\tfrac{1}{2}BC)$$
$$= 2 \times (F \times \tfrac{1}{2}BC)$$

The magnetic poles have a concave shape, so the magnetic field is radial. (See diagram of the moving coil meter on page 51 for another example of a radial field.) As the coil rotates, the **torque will be constant**, as the **force** and the **perpendicular distance** between **force** and **axis** will both be constant.

Worked example

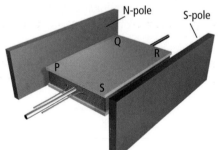

A rectangular coil *PQRS* of a model electric motor consists of 25 turns of wire with $PQ = 60\,mm$ and $QR = 30\,mm$. The magnetic induction between the poles of the magnet is 0·15T and the current in the coil is 1·8A. Calculate the torque on the coil when the coil is horizontal as shown.

$$\text{Torque on side } PQ = (F \times d) \times 25$$
$$= (BIL \times \tfrac{1}{2}QR) \times 25$$
$$= 0\text{·}15 \times 1\text{·}8 \times (60 \times 10^{-3}) \times (15 \times 10^{-3}) \times 25$$
$$= 6\text{·}075 \times 10^{-3}\,Nm$$

Torque on side $RS = 6\text{·}075 \times 10^{-3}\,Nm$
Torque on the coil $= 6\text{·}075 \times 10^{-3} \times 2 = 1.2 \times 10^{-2}\,Nm$.

This motor does not have a radial magnetic field. As the coil begins to rotate the torque will decrease as the distance between the force direction and the axis of rotation decreases.

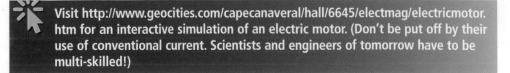

MOVING COIL AMMETER

The coil is suspended in a **radial field** and experiences a (double) **torque** when a **current flows in the coil**.

A **spring** exerts an **equal and opposite torque** on the coil, so the coil and pointer will turn so far and then stop.

The **deflection** of the coil and pointer is **proportional** to the **current** in the meter.

An expression for the torque on the coil can found. (For interest only.)

Torque $= (BI(n \times \text{Length of coil}) \, sin90° \times (\frac{1}{2} \times \text{width of coil})) \times 2$

$= BInA$, where A is the area of the coil (length × breadth).

This relationship could equally be applied to an electric motor.

current   restoring spring

MEASURING B EXPERIMENTALLY

The magnetic induction **B** between two magnets can be measured experimentally using the relationship $F = BILsin\theta$ and a current balance similar to this:

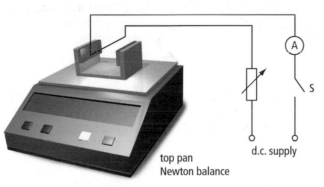

top pan
Newton balance

d.c. supply

A known length of wire is suspended rigidly between the magnets at right angles to the field. ($\theta = 90°$)

When switch **S** is closed, a **force is exerted upwards** on the wire. An **equal and opposite force** is exerted on the top pan balance **downwards**. The reading on the digital balance in **g** or **mg** is converted to **newtons**.

Typical results:

$I = 200 \, mA$

balance reading $= 25 \, mg$ $\Rightarrow F = (25 \times 10^{-3}) \times 10^{-3} \times 9{\cdot}8 \, N$

$L = 4 \, cm$

$\theta = 90°$

$B = \dfrac{F}{ILsin\theta}$

$= \dfrac{25 \times 10^{-6} \times 9{\cdot}8}{200 \times 10^{-3} \times 4 \times 10^{-2} \times 1}$

$= 3{\cdot}1 \times 10^{-2} \, T$

LET'S THINK ABOUT THIS

1   A coil of wire carrying a current in a magnetic field experiences two torques which add together. This double torque is often called a **couple**.

2   The force on one side of a coil in a magnetic field is $F = BILsin\theta \times n$, as there are **n turns** in the coil. **Don't forget to multiply by n.**

ELECTRICAL PHENOMENA

MAGNETIC FIELD AROUND A CONDUCTOR

The magnetic field around a long straight current-carrying conductor is a series of concentric circles centred on the conductor.

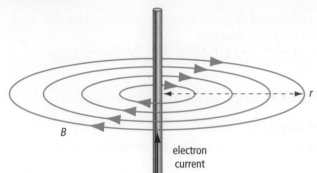

electron
current

The **magnetic induction B** at a **distance r** from the conductor is **directly proportional** to the **current I** and **inversely proportional** to the **perpendicular distance r** from the conductor.

$$B \propto \frac{I}{r}$$

$$B = \frac{\mu_0}{2\pi} \frac{I}{r}$$

The constant of proportion is $\frac{\mu_0}{2\pi}$ where μ_0 is called the **permeability of free space** and has a value $4\pi \times 10^{-7}$ TmA^{-1}. Permeability is the magnetic constant of the medium.

Worked example

A long straight conductor carries a current of $2.5\,A$. Calculate the magnetic induction at a distance of $5\,cm$ from the conductor.

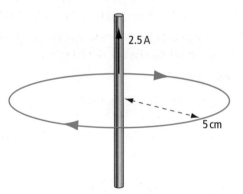

2.5 A

5 cm

$$B = \frac{\mu_0}{2\pi} \frac{I}{r}$$

$$= \frac{4 \times 3.14 \times 10^{-7} \times 2.5}{2 \times 3.14 \times 0.05}$$

$$= 1.0 \times 10^{-5}\,T$$

$$= 10\,\mu T$$

Example 1

In the above example, calculate the value of the magnetic induction at $20\,cm$ from the conductor.

Example 2

What current will give a magnetic induction of $65\,\mu T$ at a distance of $40\,mm$ from a long straight conductor?

Example 3

A long straight conductor carries a current of $8.5\,A$. At what distance from the conductor will the magnetic induction be $1.5\,mT$?

FORCE BETWEEN TWO PARALLEL CONDUCTORS

Two parallel current-carrying conductors will exert a force on each other due to one conductor being in the magnetic field of the other conductor.

Derivation of $\dfrac{F}{L} = \dfrac{\mu_0 I_1 I_2}{2\pi r}$

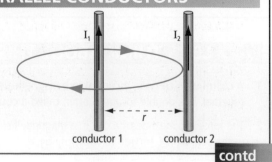

I_1

I_2

r

conductor 1   conductor 2

contd

FORCE BETWEEN TWO PARALLEL CONDUCTORS contd

Consider **two parallel conductors** distance r apart carrying currents I_1 and I_2. The **magnetic induction** at distance r from **conductor 1** is:

$$B = \frac{\mu_0}{2\pi} \frac{I_1}{r}$$

The **force** on conductor 2 due to the magnetic field from conductor 1 is:

$$F = BIL\sin\theta$$

$$= \frac{\mu_0}{2\pi} \frac{I_1}{r} \times I_2 \times L \times 1$$

The **force per unit length** on conductor 2 is:

$$\frac{F}{L} = \frac{\mu_0 I_1 I_2}{2\pi r} \ Nm^{-1}$$

The **force per unit length** on conductor 1 is also $\frac{\mu_0 I_2 I_2}{2\pi r} \ Nm^{-1}$.

> **DON'T FORGET**
>
> This derivation is examinable. The exact detail is important in derivations.

Direction of forces between parallel conductors

The direction of the force on conductor 2 can be found by using the right-hand rule.

The **electron current** in I_2 is **vertically up** and the **magnetic field** is coming **out** from the page. This will give a **force direction** on conductor 2 **towards conductor 1**. Similarly, the force on conductor 1 will be towards conductor 2. When the **currents** are in the **same direction**, the forces on the conductors will be **attractive**

When the **currents** in the conductors are in **opposite directions**, the forces on the conductors will be **repulsive**.

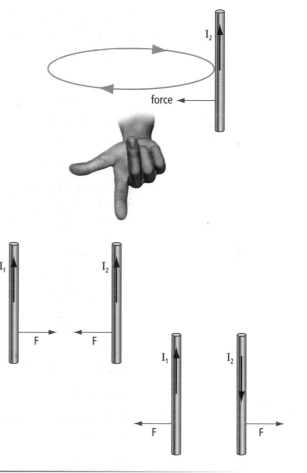

Worked example

Two long parallel conductors are 15 cm apart. The force of attraction between the conductors is $4{\cdot}5 \times 10^{-4} \ Nm^{-1}$. The current in one conductor is $8{\cdot}0$ A. Calculate the current in the other conductor.

$$\frac{F}{L} = \frac{\mu_0 I_1 I_2}{2\pi r}$$

$$4{\cdot}5 \times 10^{-4} = \frac{4\pi \times 10^{-7} \times 8 \times I_2}{2 \times \pi \times 0{\cdot}15}$$

$$I_2 = 42 \text{ A}$$

Example 4

Equal currents flow in two long straight conductors 12 cm apart. Calculate the magnitude and relative directions of the current in each conductor if the force of attraction is $25 \ \mu Nm^{-1}$.

⚙ LET'S THINK ABOUT THIS

1   The force direction between two long parallel conductors is the other way round from the more familiar rule with charges and magnets.

   same current directions $\Rightarrow$ **attractive** forces

   opposite current directions $\Rightarrow$ **repulsive** forces

2   The Scottish physicist James Clerk Maxwell discovered that electromagnetic waves had a constant equal to $\dfrac{1}{\sqrt{\varepsilon_0\mu_0}}$. Calculate the value of this constant and predict what the constant is.

MOTION IN A MAGNETIC FIELD

Previous pages have looked at the forces on current-carrying conductors. It is reasonable to assume that the force was acting on the charges which made up the current in the conductor. Now we will look at the force which acts on a single charge which moves in a magnetic field.

DERIVATION OF $F = BQV$

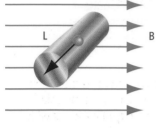

Consider a **charge** q moving with a **uniform speed** v in a conductor of **length** L at right angles to a **magnetic field** B. The charge moves a distance L in **time** t.

Substitute $L = vt$ and $I = \dfrac{q}{t}$ into $F = BIL\sin\theta$

$$F = BIL\sin\theta$$
$$= B \times \left(\frac{q}{t}\right) \times (vt) \times \sin\theta$$
$$= B \times q \times v \times \sin\theta \quad \text{consider } \theta = 90°$$
$$F = Bqv$$

The force on a **single charge** q moving with **speed** v at **right angles** to a **magnetic field** B is Bqv even when the charge moves outside the confines of a conductor.

The direction of the force can be found by using the right-hand rule. The thumb will show the force direction on a negative charge (e.g. an electron).

For the force direction on a moving positive charge, use the right-hand rule as if the moving charge is an electron and then simply take the opposite direction as given by the thumb. The following examples will illustrate this.

Worked example 1

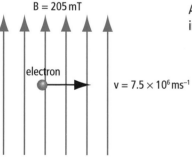

B = 205 mT

electron

v = 7.5 × 10⁶ ms⁻¹

An electron moves with a speed of 7.5×10^6 ms⁻¹ perpendicular to a magnetic field of magnetic induction $250\,mT$. Calculate the magnitude and direction of the force on the electron.

$$F = Bqv$$
$$= (250 \times 10^{-3}) \times (1.6 \times 10^{-19}) \times (7.5 \times 10^6)$$
$$= 3 \times 10^{-13}\,\text{N}$$

direction: right-hand rule: point second finger to the right (electron current)
point first finger upwards (magnetic field)
thumb points into page (force direction)

The force is 3×10^{-13} N into the page.

Worked example 2

× × × × × ×

× × × × × ×

proton

× × × × × ×

× × × × × ×

× × × × × ×

B = 480 µT (into page)

A proton moves with a speed of 1.8×10^7 ms⁻¹ perpendicular to a magnetic field of magnetic induction $480\,\mu T$. The charge on a proton is $+1.6 \times 10^{-19}\,C$.

Calculate the magnitude and direction of the force on the proton.

$$F = Bqv$$
$$= (480 \times 10^{-6}) \times (1.6 \times 10^{-19}) \times (1.8 \times 10^7)$$
$$= 1.4 \times 10^{-15}\,\text{N}$$

direction: right-hand rule: point second finger down (ignore +charge at this stage)
point first finger into page
thumb points to the left. An electron would move to the left.

The force on the proton will be 1.4×10^{-15} N **to the right**.

CIRCULAR PATH OF A CHARGED PARTICLE IN A MAGNETIC FIELD

The **force** on a **charged particle moving at right angles** to a **magnetic field** is **perpendicular** to the **direction of travel**. This will cause the charged particle to change direction only, and its speed will be unaffected.

Consider an **electron** at **point *A*** in a **magnetic field** moving to the left with **velocity *v*.** The **force *F*** on the electron will be **upwards** and have a magnitude *Bqv*.

This causes the electron to **change direction** but not its **speed**. The force on the electron is a **central force**. At **point *B*,** the force will be **to the right**.

The electron moves **clockwise** in a circle under the influence of this central force.

A **positive charge** moving in the same magnetic field will also move in a circle and it will orbit **anticlockwise** as we look at it. The central force on the positive charge will act in the opposite direction to that of the negative charge (electron). The radius of the circular orbit will not necessarily be the same as for a negative charge.

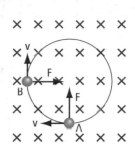

B (into page)

Radius of orbit

We can now apply Newton's Second Law to find an expression for the **radius *R*** of the circular orbit of a charged particle moving perpendicularly to a magnetic field.

unbalanced force on charged particle = **mass × acceleration**

$$Bqv = m \times \frac{v^2}{R} \quad \text{cancel } v \text{ and rearrange}$$

$$R = \frac{mv}{Bq} \quad \text{where } m = \text{the mass of the charged particle.}$$

The **radius *R*** is **directly proportional** to the **mass** and **velocity** of the charged particle and **inversely proportional** to the **magnetic induction** and **magnitude of the charge**.

Worked example

An electron of velocity $6\cdot8 \times 10^6\,\text{ms}^{-1}$ moves perpendicularly to a magnetic field of magnetic induction $0\cdot15\,\text{T}$. Calculate the radius of its orbit.

$$R = \frac{mv}{Bq}$$
$$= \frac{9\cdot11 \times 10^{-31} \times 6\cdot8 \times 10^6}{0\cdot15 \times 1\cdot6 \times 10^{-19}}$$
$$= 2\cdot6 \times 10^{-4}\,\text{m}.$$

A proton with the same velocity as the electron moving in the same magnetic field will have a much bigger orbit radius, as the mass of the proton is bigger than the mass of the electron. (*B*, *q* and *v* would be unchanged.)

Bubble chamber

The paths taken by charged particles can be seen using a bubble chamber containing liquid hydrogen in a magnetic field. The charged particles leave an ionisation track, and tiny bubbles form around the ions. These tracks can then be photographed for further study.

This photograph shows charged particles moving in a circle and spiralling inwards as they lose energy. (The radius decreases as the velocity decreases.) The particles can be identified using the magnitude and direction of the magnetic field as well as measurements of the initial radius.

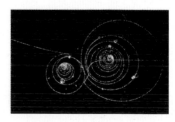

LET'S THINK ABOUT THIS

1 Practise using the right-hand rule. You will be expected to predict the direction of the force on a charged particle moving in a magnetic field.

2 The relationship $R = \frac{mv}{Bq}$ is not in the data booklet, but the physics leading to it is worth remembering.

HELICAL MOTION IN A MAGNETIC FIELD

A charged particle entering a uniform magnetic field at 90° to the field lines will follow a circular path as previously described. If the particle enters the magnetic field at **angle** θ (0° < θ < 90°), only a **component of the particle's velocity** will be **perpendicular** to the **magnetic field lines**.

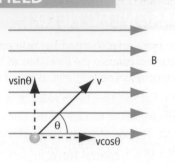

The component of **v** perpendicular to **B** is $v_\perp = v\sin\theta$

The component $v_\perp$ will result in a force on the charged particle, causing it to change direction and move in a circle with a constant speed of $v\sin\theta$.

The component of **v** parallel to **B** is $v_\parallel = v\cos\theta$

There is no force on the charged particle parallel to the magnetic field, so $v_\parallel$ will remain constant.

The charged particle will move in a circle perpendicular to **B** as well as moving in a straight line parallel to **B**. This will result in the particle following a **helical path** as shown. The axis of the helix will be in the direction of the magnetic field.

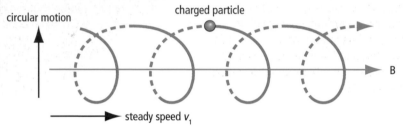

Looking from the left-hand side at the helical motion, you should see that the charged particle is moving clockwise. Use the right-hand rule to confirm that the charge on the particle must be negative. Positive charged particles would rotate anticlockwise as seen from the left.

Numerical calculations involving the helical path can be made (e.g. the radius of the helix), but care must be taken to use only the components of the particle's velocity and not the initial speed.

Worked example

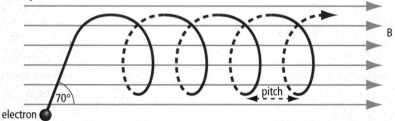

An electron travelling at $2.3 \times 10^7 \, ms^{-1}$ enters a uniform magnetic field at an angle of 70° as shown. The magnetic induction $= 0.18 \, T$.

Calculate the radius of the helix.

$$v_\perp = v\sin70 = 2.3 \times 10^7 \times \sin70 = 2.16 \times 10^7 \, ms^{-1}$$

$$\text{Force on electron} = Bq(v\sin\theta) = 0.18 \times (1.6 \times 10^{-19}) \times (2.16 \times 10^7)$$

$$= 6.22 \times 10^{-13} \, N$$

$$F = m \times \frac{v^2}{R} \Rightarrow R = m \times \frac{v^2}{F} = \frac{9.11 \times 10^{-31} \times (2.16 \times 10^7)^2}{6.22 \times 10^{-13}} = 6.8 \times 10^{-4} \, m.$$

Calculate the pitch of the helix (i.e. the distance between adjacent loops)

$$v_\parallel = v\cos70° = 2.3 \times 10^7 \times \cos70° = 7.87 \times 10^6 \, ms^{-1}$$

$$\text{time for 1 revolution} = \frac{2\pi R}{v_\perp} = \frac{2 \times 3.14 \times 6.8 \times 10^{-4}}{2.16 \times 10^7} = 1.98 \times 10^{-10} \, s.$$

$$\text{pitch} = v_\parallel \times \text{time or 1 revolution} = (7.87 \times 10^6) \times (1.98 \times 10^{-10}) = 1.6 \times 10^{-3} \, m.$$

contd

HELICAL MOTION IN A MAGNETIC FIELD contd

Aurora Borealis (Northern Lights)

Charged particles emitted by the Sun and travelling towards the Earth will first enter the Earth's magnetic field. Unless θ is 0° or 90°, they will change direction and spiral along the Earth's magnetic field lines.

When the charged particles reach the Earth's atmosphere, they collide with air atoms and molecules producing light. Collisions with atomic oxygen produce green light while collisions with nitrogen produce pink light. A complete band of coloured light can be seen from space during periods of strong sunspot activity. From the Earth, an observer will see part of this band – the Aurora Borealis or Northern Lights.

This light display is regularly seen in the north of Scotland and occasionally in the south of Scotland well away from light pollution. Thanks go to Joshua Strang for the use of this photograph, taken in Alaska in 2005. Similar aurora can be seen near the South Pole, called the Aurora Australis.

⚙ LET'S THINK ABOUT THIS

The helical pattern can be explained by Newton's laws. Newton II applies for motion perpendicular to the magnetic field (unbalanced force causing centripetal acceleration). Newton I applies for motion parallel to the magnetic field (steady speed).

ELECTRICAL PHENOMENA

COMBINING ELECTRIC AND MAGNETIC FIELDS

A charged particle moving in a combined electric and magnetic field will experience two forces:

- $F_B = Bqv_\perp$ due to the magnetic field
- $F_E = Eq$ due to the electric field.

If the magnitudes of these two forces are equal but their directions are opposite, then the resultant force on the charged particle will be zero and it will move with a steady speed (Newton I).

The electric and magnetic fields must be mutually perpendicular to each other and are often called "crossed" electric and magnetic fields.

The diagrams show one way of doing this.

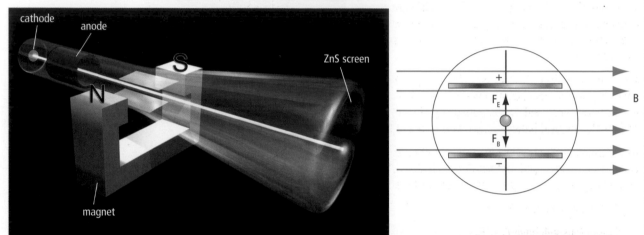

The permanent magnets provide the magnetic field **B**, and the parallel plates inside the evacuated tube provide the electric field **E**.

Consider an electron moving between the plates. The end view shows the electron coming straight towards you with the two vertical forces F_E and F_B (the weight of the electron is negligible).

If $F_B = F_E$

$$Eq = Bqv$$

$$v = \frac{E}{B}$$

An electron with a speed of $\frac{E}{B}$ will move between the plates in a straight line.

J.J. Thomson

In 1896, J.J. Thomson used an evacuated tube similar to the above tube to measure $\frac{q}{m}$, the charge-to-mass ratio of the negative "rays" coming from the cathode in the evacuated tube.

An expression for $\frac{q}{m}$ can be found as follows:

$$F_E = F_B$$

$$Eq = Bqv = \frac{mv^2}{R}$$

$$\frac{q}{m} = \frac{v}{BR}$$

$$= \frac{\frac{E}{B}}{BR}$$

$$\frac{q}{m} = \frac{E}{B^2 \times R} \qquad E = \frac{voltage\ across\ the\ plates}{plate\ separation}$$

E and **B** could be measured, and **R** is found by switching off the plate voltage and taking measurements of the curvature of the subsequent path.

contd

COMBINING ELECTRIC AND MAGNETIC FIELDS contd

Thomson found that the value for $\frac{q}{m}$ was constant regardless of the cathode material, and the mass must be very small. He suggested that he was dealing with a negative sub-atomic particle common to all matter. He proposed naming it a "corpuscle", but this name didn't catch on. The word "electron" was proposed by the physicist George Stoney, who combined the word "electric" with the suffix "-on".

J.J. Thomson was awarded the Nobel Prize in 1906 for the discovery of the electron.

Visit http://www.dnatube.com/video/1816/The-Discovery-of-the-Electron-2-of-15 for a short film clip about J.J. Thomson.

Mass spectrometer

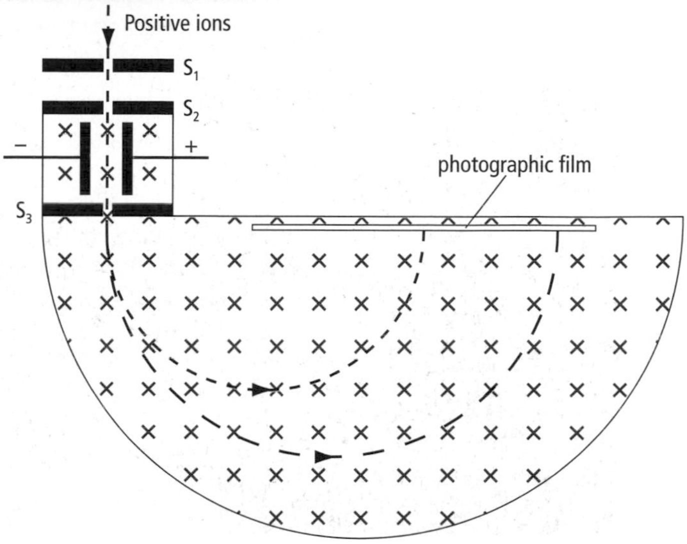

A sample of various charged particles of different speeds, masses and charges can be separated and analysed using a **mass spectrometer**. The diagram shows one type of mass spectrometer.

A stream of **positive ions** passes through slits S_1 and S_2, emerging as a narrow beam of various masses and speeds. The region between S_2 and S_3 has a magnetic field into the page and an electric field between the plates (crossed fields).

As described on the previous page, ions with a speed of $v = \frac{E}{B}$ will proceed undeviated and pass through slit S_3. Ions with a different speed will change direction and not reach slit S_3.

Faster ions will have $F_B > F_E$

Slower ions will have $F_B < F_E$

The ions which pass through slit S_3 leave the electric field and come under the influence of the magnetic field only. They will then move in semicircles before hitting a photographic film or other sensor which records the event. The **radius** of the semicircle is **proportional to the mass**, so different masses will move in different semicircles.

$$R = \frac{mv}{Bq} \qquad v, B \text{ and } q \text{ are constant } (q \text{ is usually } +1\cdot6 \times 10^{-19}\,C).$$

Once R is measured, the value of m can be calculated.

Worked example

Use this information about the mass spectrometer above to calculate the mass of the bigger of the two ions separated out.

$$p.d. \text{ across the plates } = 2\,kV$$

$$\text{separation of the plates } = 50\,mm$$

$$\text{magnetic induction } B = 0.45\,T$$

$$\text{diameter of larger semicircle } = 94\,mm$$

$$\text{charge on ion } = (+)1.6 \times 10^{-19}\,C$$

$$\text{electric field strength } E = \frac{V}{d} = \frac{2000}{50 \times 10^{-3}} = 4.0 \times 10^4\,NC^{-1}$$

$$\text{velocity of ion } = \frac{E}{B} = \frac{4 \times 10^4}{0\cdot45} = 8\cdot89 \times 10^4\,ms^{-1}$$

$$\text{mass of ion } = \frac{RBq}{v} = \frac{47 \times 10^{-3} \times 0.45 \times 1\cdot6 \times 10^{-19}}{8\cdot89 \times 10^4} = 3\cdot8 \times 10^{-26}\,kg.$$

LET'S THINK ABOUT THIS

The magnetic field around the evacuated tube is often provided by two electromagnets called **Helmholtz coils**.

SELF-INDUCTANCE

Induced e.m.f.

We have already seen in earlier studies that moving a permanent magnet inside a coil of wire causes a reading on the meter. A **voltage** is **induced** in the coil.

Any **changing magnetic field** will **induce** an electromotive force (e.m.f.) in a **conductor** placed in the **magnetic field**.

Another example of a changing magnetic field is in the region around a current-carrying conductor **when the current is changing**. An increasing current will cause the magnetic field at a particular point to increase in value.

2.5 A

10 µT

5.0 A

20 µT

The magnetic induction **B** has a value of 10 µ*T* at a distance of 5 cm from a conductor carrying a current of 2.5 A.

If the current increases to 5.0 A, the value of **B** at the same position will increase to 20 µ*T*.

The magnetic field at all points around the conductor will increase as the current increases.

But the conductor itself is in its own changing magnetic field, so a **self-induced e.m.f.** will be produced across the conductor. This will be in addition to the potential difference causing the current in the first place.

Changing currents and changing magnetic fields will occur:

● in d.c. circuits when switching the circuit on or off

● in a.c. circuits all the time.

Inductors

A coil of wire has a greater magnetic field around it than a straight length of wire carrying the same current. The self-induced e.m.f. generated by a coil when its magnetic field changes can be quite considerable.

A coil of wire is often called an **inductor** because of its ability to induce an e.m.f. across itself. Symbols for an inductor are:

air-cored inductor   *iron-cored inductor*

INDUCTOR IN A D.C. CIRCUIT

Connecting an inductor into a series d.c. circuit illustrates well the effects of the self-induced e.m.f.. The inductor is assumed to have negligible resistance.

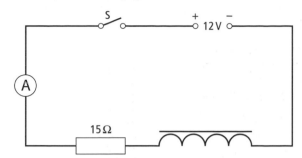

Switch **S** is closed for several seconds then opened again. The graph shows how the current changes during this time.

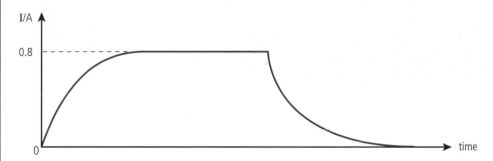

The current does not reach its maximum value immediately. This is because the increasing current causes a changing magnetic field around the inductor. An e.m.f. is induced in the conductor, and this opposes the build-up of current. The induced e.m.f. is often called a **back e.m.f.**, as it **opposes the change in current**. The **back e.m.f. reduces to zero** as the **current approaches its maximum value**.

The maximum current is found by Ohm's law: $I = \frac{V}{R} = \frac{12}{15} = 0.8\,A$. There is a magnetic field around the inductor, but it is not changing, so the back e.m.f. is zero at maximum current.

On opening the switch, the current reduces, and this causes a changing (collapsing) magnetic field. A back e.m.f. is induced in the inductor, opposing the decreasing current. The decreasing current lasts for a short time after the switch is opened.

An inductor with **more turns** will create a **bigger back e.m.f.** and the current will take even **longer** to reach its maximum value.

An inductor with **fewer turns** will create a **smaller back e.m.f.** and the current will take **less time** to reach its maximum value.

Notice the **maximum current will be unchanged** if the inductors have negligible resistance.

(A more exact way of defining the size of an inductor will follow on page 62.)

Energy considerations

If the induced e.m.f. did not oppose the build-up of current but instead acted the opposite way, then the current increase would be even greater, so producing a bigger induced e.m.f.. This constantly increasing cycle would create its own energy once started. The conservation of energy does not allow this, so the induced e.m.f. in an inductor must oppose the changing current.

Lenz's Law states that the **induced e.m.f. opposes** the **change in current** causing it.

LET'S THINK ABOUT THIS

Any explanation of self-inductance must include details about the changing magnetic field.

INDUCTANCE

The previous page discussed how a self-induced e.m.f. is produced in a coil of wire whenever the current in the coil changes. We also saw evidence that the self-induced e.m.f. opposes the change in current. Mathematically, the relationship is direct proportion.

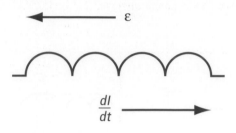

$\varepsilon \propto -\dfrac{dI}{dt}$ where ε is the **self-induced e.m.f.** in a coil or inductor.

The **negative sign** shows that ε is in the **opposite direction to** $\dfrac{dI}{dt}$.

Introducing the **constant of proportion**, we get:

$\varepsilon = -L\dfrac{dI}{dt}$ where **L** is the **self-inductance** or **inductance** of the coil.

The unit of L is $\dfrac{V}{As^{-1}} = VsA^{-1} = \textbf{H}$ (henry)

A coil has an **inductance of 1 henry** when an **e.m.f. of 1 volt** is **self-induced** when the **current changes** at a rate of **1 ampere per second**.

An inductor with an inductance of 2·5 H will induce a back e.m.f. of 2·5 V when the rate of change of current is 1 As⁻¹.

Worked example

Calculate the self-induced e.m.f. in an inductor of inductance 0·75 H when the rate of change of current is 3·5 As⁻¹.

$\varepsilon = -L\dfrac{dI}{dt}$

$= -0·75 \times 3·5$

$= -2·6\,V$

DON'T FORGET

ε has a **negative** sign. The minus sign must be included when substituting for ε or back e.m.f..

Worked example

A back e.m.f. of 8V is induced in a coil when the current changes at 2 As^{-1}. Calculate the inductance of the coil.

$\varepsilon = -L\dfrac{dI}{dt}$

$-8 = -L \times 2$ or $L = -\dfrac{\varepsilon}{\frac{dI}{dt}} = -\dfrac{(-8)}{2} = 4\,H$

$L = 4\,H$

DON'T FORGET

It is incorrect physics to simply drop the minus sign to get a positive value for L.

Energy stored in an inductor

An inductor **stores energy** in the magnetic field around it. The relationship for the energy **E** stored in an inductor is:

$E = \dfrac{1}{2}LI^2$ where **L** is the **inductance** of the inductor and **I** is the **current** in the inductor.

Example 1

Calculate the energy stored in a 4·0 H inductor carrying a current of 600 mA.

Example 2

A 50mH inductor stores 160 μJ of energy. Calculate the current flowing in the inductor.

DON'T FORGET

Remember to **square the current** during calculations for energy – a common oversight.

contd

INDUCTANCE contd

Extended worked example

The following is an example of an exam-type question of an inductor in a d.c. circuit. The circuit shows an inductor of negligible resistance and a battery of negligible internal resistance.

The graph shows the growth in the current after switch S is closed.

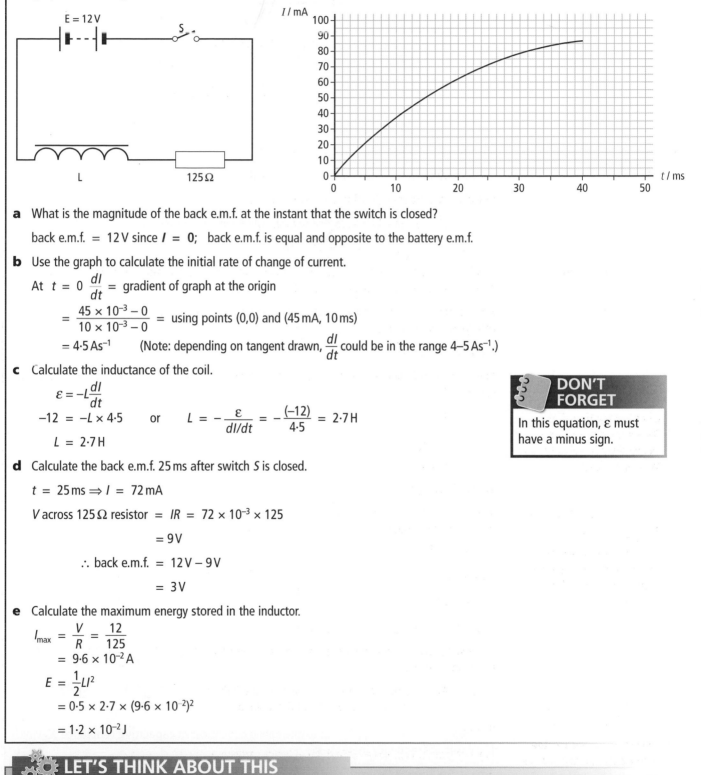

a What is the magnitude of the back e.m.f. at the instant that the switch is closed?

back e.m.f. $= 12\,V$ since $I = 0$; back e.m.f. is equal and opposite to the battery e.m.f.

b Use the graph to calculate the initial rate of change of current.

At $t = 0$ $\dfrac{dI}{dt} =$ gradient of graph at the origin

$= \dfrac{45 \times 10^{-3} - 0}{10 \times 10^{-3} - 0} =$ using points (0,0) and (45 mA, 10 ms)

$= 4\cdot5\,As^{-1}$ (Note: depending on tangent drawn, $\dfrac{dI}{dt}$ could be in the range 4–5 As^{-1}.)

c Calculate the inductance of the coil.

$\varepsilon = -L\dfrac{dI}{dt}$

$-12 = -L \times 4\cdot5$ or $L = -\dfrac{\varepsilon}{dI/dt} = -\dfrac{(-12)}{4\cdot5} = 2\cdot7\,H$

$L = 2\cdot7\,H$

> ### DON'T FORGET
>
> In this equation, ε must have a minus sign.

d Calculate the back e.m.f. 25 ms after switch S is closed.

$t = 25\,ms \Rightarrow I = 72\,mA$

V across $125\,\Omega$ resistor $= IR = 72 \times 10^{-3} \times 125$

$= 9\,V$

$\therefore$ back e.m.f. $= 12\,V - 9\,V$

$= 3\,V$

e Calculate the maximum energy stored in the inductor.

$I_{max} = \dfrac{V}{R} = \dfrac{12}{125}$

$= 9\cdot6 \times 10^{-2}\,A$

$E = \dfrac{1}{2}LI^2$

$= 0\cdot5 \times 2\cdot7 \times (9\cdot6 \times 10^{-2})^2$

$= 1\cdot2 \times 10^{-2}\,J$

⚙ LET'S THINK ABOUT THIS

1 The minus sign must be included in the equation $\varepsilon = -L\dfrac{dI}{dt}$

2 If calculating L or $\dfrac{dI}{dt}$, you must substitute a negative value for the back e.m.f. ε.

ELECTRICAL PHENOMENA

INDUCTORS IN A.C. CIRCUITS

In an **a.c. circuit**, the **current** and its associated **magnetic field** are continually changing. An inductor in an a.c. circuit will **always self-induce a back e.m.f.** and not just when switching on or off as was the case in a d.c. circuit.

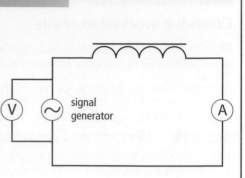

As the **frequency** of the a.c. **increases**, the **rate of change of current will increase** and we would expect the **back e.m.f.** across an inductor to **increase**. The following experiment looks at how the current varies as the frequency is increased in an inductive circuit.

A signal generator is used as the a.c. power supply. This allows the frequency of the alternating current to be varied, and the frequency can be read from the dial on the signal generator. It is important to keep the output voltage of the supply constant, so an a.c. voltmeter is connected across the output terminals of the signal generator. An a.c. ammeter takes readings of the r.m.s. current in the circuit for various frequencies set by the signal generator, and a graph of **current I** against **frequency f** is drawn.

The graph shows an inverse-type relationship between I and f. As the **frequency increases**, the r.m.s. **current** in the circuit **decreases**.

To investigate further what type of relationship exists between **current I** and **frequency f**, we should look for some combination of the two variables which gives a constant. For inverse relationships, multiply the variables together. Record this in an extra column in the results table.

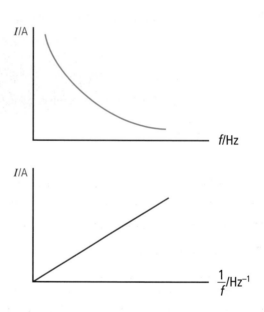

f	I	$f \times I$

If $f \times I$ is a constant, then:

$$f \times I = k$$
$$I = \frac{k}{f}$$
$$I \propto \frac{1}{f} \qquad k \text{ is constant of proportion.}$$

Alternatively, plot a graph of I against $\frac{1}{f}$ and look for a straight line through the origin as proof of $I \propto \frac{1}{f}$.

The graphical method is more time-consuming than the search for the constant of proportion.

The opposition of an inductor to alternating current is called **inductive reactance X_L ohms**.

X_L is proportional to the a.c. frequency, so the **current decreases** as the **frequency increases**:

$$X_L \propto f$$

Capacitors also show an opposition to alternating current called **capacitive reactance X_C** which **decreases** as **frequency increases**:

$$X_C \propto \frac{1}{f}$$

contd

64

INDUCTORS IN A.C. CIRCUITS contd

Uses of inductors

The transformer has been studied in some detail in earlier physics courses. The **changing magnetic field** in the **primary coil induces** a **voltage** in the **secondary coil**. The transformer can step up or down an a.c. voltage.

A mobile-phone charger is really a step-down transformer (230 V to 5 V) and also a rectifier to change a.c. into d.c.

Other uses of inductors include **chokes** and **induction loops**. **Chokes** are used in analogue audio systems to filter out high-frequency electrical interference signals. Car audio systems must filter out all the interference from the spark plugs and car engine electrical systems before the audio signal reaches the loudspeaker.

An **induction loop** enables hearing-aid users to hear clearly speech or music in theatres, churches etc. Current from the microphone is fed into a loop of wire going round the perimeter of the room. The magnetic field is modulated by the speech and reaches all parts within the loop.

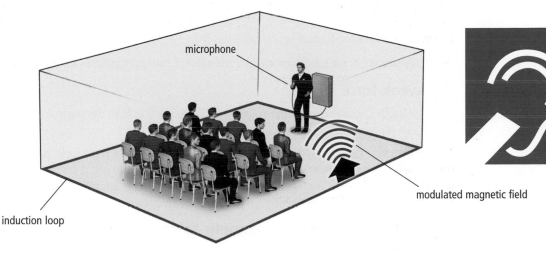

microphone

induction loop

modulated magnetic field

The changing magnetic field induces an audio frequency voltage in a tiny coil in the hearing aid, and the person can hear everything clearly even when moving around.

Look out for the sign that an induction loop has been fitted.

Traffic-light cameras

Traffic-light speed cameras are used in conjunction with two induction loops buried under the road junction. A metal car above a buried coil will act like an iron core and increase the inductance of the coil. If this happens in the first coil and then the second coil when the lights are at red, a picture is taken.

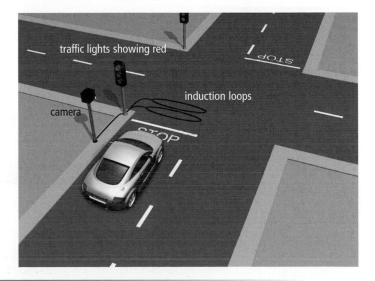

traffic lights showing red

induction loops

camera

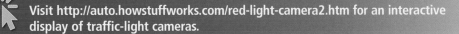

Visit http://auto.howstuffworks.com/red-light-camera2.htm for an interactive display of traffic-light cameras.

FORCES OF NATURE

Physicists have identified four fundamental forces of nature:

- the strong force
- the weak force
- the electromagnetic force
- the gravitational force.

We are already familiar with the final two forces in our everyday lives but less familiar with the first two, which exist only inside the nucleus of an atom.

The strong force

The strong force holds all the protons and neutrons together inside the nucleus. The strong force overcomes the repulsive forces between positive protons and holds the nucleus together.

The range of the strong force is very short – less than 10^{-14} m.

As the name implies, the strong force is the strongest of the four fundamental forces.

The weak force

The weak force is responsible for the emission of a beta particle from the nucleus of a radioactive element.

A **neutron** inside the nucleus **decays** into a **proton** and an **electron** (the **beta particle**). The weak force overcomes the attractive force between the beta particle and a proton and ejects the beta particle from the nucleus. The weak force is a short-range force found only inside the nucleus.

DON'T FORGET

The beta particle is an electron which originates in the nucleus. It is not an orbiting electron.

Gravitational and electromagnetic forces

We have already seen that gravitational and electromagnetic forces extend over great distances.

The following example compares these two forces between the proton and orbiting electron in a hydrogen atom where the orbit radius is $5{\cdot}3 \times 10^{-11}$ m.

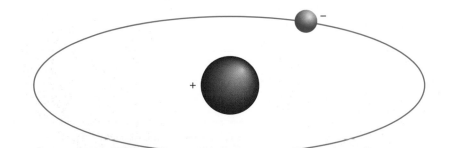

Gravitational force	Electromagnetic force
$F = \dfrac{Gm_p m_e}{r^2}$ $= \dfrac{6{\cdot}67 \times 10^{-11}\,(1{\cdot}67 \times 10^{-27}) \times (9{\cdot}11 \times 10^{-31})}{(5{\cdot}3 \times 10^{-11})^2}$ $= 3{\cdot}6 \times 10^{-47}\,\text{N}$	$F = \dfrac{1}{4\pi\varepsilon_0}\dfrac{Q_p Q_e}{r^2}$ $= 9 \times 10^9 \times \dfrac{(1{\cdot}6 \times 10^{-19}) \times (1{\cdot}6 \times 10^{-19})}{(5{\cdot}3 \times 10^{-11})^2}$ $= 8{\cdot}2 \times 10^{-8}\,\text{N}$

The electromagnetic force is greater than the gravitational force by a factor of 10^{39}, so the gravitational force is negligible in this case.

QUARKS

Nuclear physicists have identified many short-lived sub-atomic particles in addition to protons, neutrons and electrons. **Protons** and **neutrons** are now known to be made up of **quarks**.

Quarks were first postulated in 1964 by both Murray Gell-Mann and George Zweig. They proposed the existence of three quarks called **up**, **down** and **strange**, and in the next 30 years these were identified experimentally along with another three – **top**, **bottom** and **charm**.

Quarks have fractional electric charge values: either $-\frac{1}{3}\mathbf{e}$ or $+\frac{2}{3}\mathbf{e}$ (see table).

Flavour	Charge
Up	$+\frac{1}{3}e$
Down	$-\frac{1}{3}e$
Charm	$+\frac{2}{3}e$
Strange	$-\frac{1}{3}e$
Top	$+\frac{2}{3}e$
Bottom	$-\frac{1}{3}e$

Protons and neutrons are now known to consist of three quarks each.

A proton consists of two up quarks and one down quark.

A neutron consists of one up quark and two down quarks.

The charge on a proton is +1e since $+\frac{2}{3}e + \frac{2}{3}e + (-\frac{1}{3}e) = +1e$

The charge on a neutron is zero since $+\frac{2}{3}e + (-\frac{1}{3}e) + (-\frac{1}{3}e) = 0$

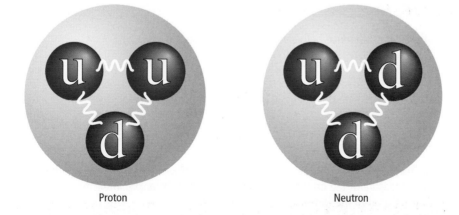

Proton   Neutron

The etymology of the physics name "quark" is interesting. Gell-Mann suggested "quark", as he could remember a reference to the term "three quarks" in the novel *Finnegans Wake* by James Joyce, and the number three was prominent in his theory.

Zweig, working independently, suggested the name "ace" for these particles. The term "quark" prevailed.

One final point: quark is usually pronounced "quork", as in "York".

Enter the word "quark" into a search engine for more information.

WAVES

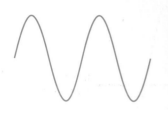

A wave is a disturbance which moves from one place to another in a medium **transferring energy** as it moves. Particles in the medium can be displaced as the wave passes but return to their undisturbed positions once the wave has passed through. There is **no net mass transport** in wave motion. Only energy is transferred.

The simplest mathematical model of a wave is based on the **sine** or **cosine** function. The diagram shows a **sine** wave which begins at **zero displacement**.

The cosine wave begins at **maximum positive displacement** on a crest.

Phase difference

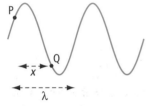

While you are already familiar with the terms **wavelength**, **frequency** and **amplitude** of a wave, the concept of **phase difference** is new.

Consider two points **P** and **Q** on a sine wave which are a distance **x** apart. The phase difference is a measure of the **separation** of these two points as a **fraction** of the wavelength and expressed as an **angle** in **radians**.

The phase difference $\Phi = \dfrac{x}{\lambda} \times 2\pi$

Points which are $\dfrac{1}{2}\lambda$ apart will have a phase difference of π.

Worked example

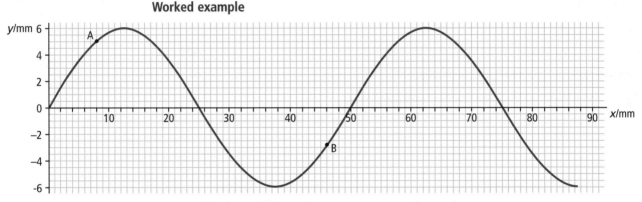

Calculate the phase difference between points A and B on the wave.

$$\Phi = \frac{x}{\lambda} \times 2\pi$$

$$= \frac{(46 - 8)}{50} \times 2 \times 3{\cdot}14$$

$$= 4{\cdot}8 \, \text{rad}$$

Intensity of a wave

As the **amplitude** of a wave **increases**, the **energy transferred** by the wave also **increases**. **Intensity** is a measure of the **energy per second per unit area**. The intensity of a wave is directly **proportional** to the **(amplitude)²** of the wave.

If the **amplitude** of a wave **doubles**, the **intensity** of the wave increases **four times**.

A useful formula for this direct-proportion relationship is:

$$\frac{I_1}{(A_1)^2} = \frac{I_2}{(A_2)^2} \qquad (= \text{constant of proportion})$$

Example 1

How does the amplitude change when the intensity of a wave doubles?

Example 2

How does the intensity of a wave change when the amplitude halves?

Equation of a travelling wave

A travelling wave can be generated on an elasticated rope by repeatedly moving one of its ends up and down.

The wave moves to the right, and each point on the rope changes position with time. The diagram shows the profile of the wave at one instant in time.

The displacement y of a point on the rope located at the origin is given by the expression:

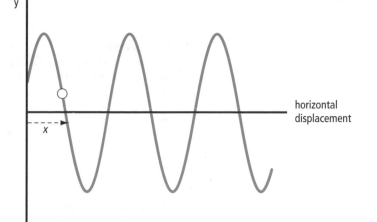

$$y = A\sin\omega t$$

$$= A\sin 2\pi f t \qquad \text{where } \boldsymbol{f} \text{ is the wave } \textbf{frequency} \text{ and } \boldsymbol{A} \text{ is the wave } \textbf{amplitude}.$$

At a horizontal distance +x from the origin, the displacement of a rope particle is given by:

$$y = A\sin(2\pi f t - \frac{x}{\lambda} \times 2\pi)$$

$$y = A\sin 2\pi(ft - \frac{x}{\lambda})$$

The equation of an identical wave travelling to the left is given by:

$$y = A\sin 2\pi(ft + \frac{x}{\lambda})$$

DON'T FORGET

Moving $\sin\theta$ to the right by a becomes $\sin(\theta - a)$. You should be familiar with this from your studies of maths.

Worked example

A travelling wave has the equation $y = 0.65\sin(5t - 3x)$ where x and y are in metres and t in s.

By inspection: the amplitude $= 0.65\,\text{m}$.

By calculation: $2\pi f = 5$ (comparing coefficients of t)

$$f = \frac{5}{2 \times 3.14} = 1.27\,\text{Hz}$$

Also $\dfrac{2\pi}{\lambda} = -3$

$$\lambda = \frac{2 \times \pi}{3} = 2.1\,\text{m}.$$

The speed of the wave is calculated: $v = f \times \lambda = 1.27 \times 2.1 = 2.7\,\text{ms}^{-1}$.

Example 3

The equation of a wave is $y = 2.5 \times 10^{-3} \times \sin(45t + 0.7x)$ x, y in m and t in s.

Calculate the amplitude, frequency, wavelength, speed and direction of the wave.

Example 4

Write down the equation of a wave travelling to the right with a speed of $0.75\,\text{ms}^{-1}$, wavelength $2.5\,\text{cm}$ and amplitude of $4\,\text{cm}$.

⚙ LET'S THINK ABOUT THIS

The equation of a travelling wave $y = A\sin 2\pi(ft - \frac{x}{\lambda})$ can be rewritten to include wave speed v and period T.

$$y = A\sin 2\pi(\frac{t}{T} - \frac{x}{\lambda})$$

$$y = A\sin 2\pi f(t - \frac{x}{v})$$

Show that these three equations are equivalent.

The sine of an angle has no units or dimensions. Show that this is the case with each of these three expressions for a travelling wave.

STATIONARY WAVES

A travelling wave moving to the right along a stretched slinky can be reflected if the other end of the slinky is fixed.

The incident wave and the reflected wave will interfere, and a stationary wave is set up at various wave frequencies.

One such stationary wave looks like this:

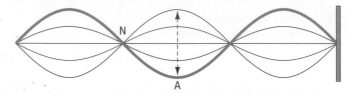

Some points on the stationary wave do not move. These points are called **nodes** (**N**). Nodes are positions of **zero disturbance** in a stationary wave. The diagram shows four nodes. **Halfway** between the nodes are points of **maximum disturbance** called **antinodes** (**A**). This diagram shows three antinodes.

Note also that the **distance between nodes** equals **half a wavelength**.

Microwave stationary waves

Microwaves are transmitted towards a metal plate which reflects the waves back towards the transmitter. A stationary wave is produced between the incident and reflected waves. The location of nodes can be found using a microwave detector connected to an ammeter.

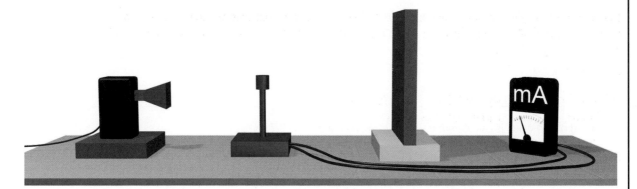

DON'T FORGET

N N N N N N
← - - - - - - →
8.4 cm

Divide by 6, not 7.

The ammeter will show a **minimum** reading when the detector is at a **node**. Note the positions of two adjacent nodes. The distance between two adjacent nodes is equal to $\frac{1}{2}\lambda$.

Typical results:

distance between 7 consecutive nodes = 8.4 cm

distance between 2 consecutive nodes = 1.4 cm

$$\frac{\lambda}{2} = 1.4$$

$$\lambda = 2.8 \text{ cm}$$

DON'T FORGET

Microwaves travel at the speed of light.

The frequency of the microwave transmitter can now be calculated and compared to the value given by the manufacturer.

$$v = f \times \lambda$$

$$f = \frac{v}{\lambda} = \frac{3 \times 10^8}{2.8 \times 10^{-2}} = 10.7 \text{ GHz}$$

contd

STATIONARY WAVES contd

Stationary sound waves

Sound from a loudspeaker can be sent along a closed tube, and the **reflected** sound wave will **interfere** with the **incident** wave to set up a **stationary** wave at various **frequencies** or **lengths of tube**. The frequency of the sound is adjusted using the dial on the signal generator until a loud resonant sound is heard coming from the tube. A **stationary sound wave** has been formed in the tube. The stationary sound wave is often represented as a **transverse** wave with **nodes** (**N**) and **antinodes** (**A**). Notice that a **node is formed** at the **closed** end of the tube.

One method of locating the nodes is to place some fine powder along the base of the tube.

Switching on the signal generator and adjusting the frequency until a resonant sound is heard causes the layer of powder to change shape as follows:

The powder will experience forces due to the disturbance of the air molecules at and near the antinodes. Powder at the nodes will not be disturbed, and the amount of powder here will build up.

Example 1

Calculate the speed of sound using the following results:

distance between adjacent nodes = 92 mm

sound frequency = 1750 Hz

Example 2

Small mounds of powder are formed 12 cm apart in the tube above. If the speed of sound is 340 ms^{-1}, calculate the frequency of the sound.

🛠 LET'S THINK ABOUT THIS

The standing wave is caused by the **superposition** (or overlapping) of the incident wave and the reflected wave.

Mathematically, the displacement y will be:

$$y = A\sin 2\pi(ft - \frac{x}{\lambda}) + A\sin 2\pi(ft + \frac{x}{\lambda})$$

$$= 2A\sin(2\pi ft)\cos(2\pi\frac{x}{\lambda}) \text{ using a trig addition relationship.}$$

When $x = 0, \frac{1}{2}\lambda, \lambda, \frac{3}{2}\lambda, \ldots$ the value of y is a maximum since $\cos(2\pi\frac{x}{\lambda})$ will be ±1. Antinodes occur every half-wavelength. Nodes which are midway between antinodes also occur every half-wavelength.

Note that the amplitude of the standing wave is $2A\cos(2\pi\frac{x}{\lambda})$ and is dependent on the value of x.

DOPPLER EFFECT

We are familiar with the Doppler effect from listening to the change in pitch from a fire engine's siren as it passes at speed. The Doppler effect is the **change in frequency** observed when a **source of sound waves** is moving **relative** to an observer.

Moving source

An ambulance with its siren on approaches a stationary observer. The expression for the frequency heard by the observer is derived as follows:

The speed of sound in air is v (340 ms⁻¹); the speed of the ambulance (source) is v_s; the siren frequency is f_s.

The stationary ambulance sends out f_s waves in 1 second occupying a distance v (as v is the **distance** travelled per second).

The moving ambulance causes these f_s waves in 1 second to be squeezed into a distance of $v - v_s$.

The reduced wavelength λ' is:

$$\lambda' = \frac{distance}{number\ of\ waves} = \frac{v - v_s}{fs}$$

The frequency f heard by the observer is:

$$f = \frac{v}{\lambda'} = \frac{v}{\frac{v - v_s}{fs}} = \frac{v}{(v - v_s)} f_s$$

DON'T FORGET

This derivation considers what happens in one second.

If the ambulance is moving away from the observer, then f waves will occupy a distance $v + v_s$ in 1 second; and, following the same procedure, the frequency heard is $f = \frac{v}{(v + v_s)} f_s$

Worked example

An ambulance travelling at 25 ms⁻¹ emits a sound of frequency 900 Hz as it approaches a stationary observer.

a Calculate the frequency heard by the observer.

$$f = \frac{v}{(v \pm v_s)} f_s = \frac{v}{(v - v_s)} f_s = \frac{340}{(340 - 25)} \times 900 = 971\ Hz$$

b Calculate the frequency heard by the observer after the ambulance has passed.

$$f = \frac{v}{(v + v_s)} f_s = \frac{340}{(340 + 25)} \times 900 = 838\ Hz.$$

Note that the observed frequencies are not symmetrical about the stationary frequency.

DON'T FORGET

The data book combines both expressions for a moving source, giving $f = \frac{v}{(v \pm v_s)} f_s$.
Use the "−" option when the source approaches the observer. This decreases the denominator, increasing f.

Example 1

A police car's siren has a frequency of 1200 Hz. A stationary observer hears the frequency as 1250 Hz.

Calculate the speed of the police car and state whether it is approaching or moving away from the observer.

Example 2

An observer hears a sound of frequency 1600 Hz from a car moving away from her at 30 ms⁻¹.

Calculate the actual frequency of the sound emitted from the car.

contd

DOPPLER EFFECT contd

Moving observer

A moving observer approaching a stationary source of sound will also hear the Doppler effect.

An observer moves with speed v_o towards an alarm bell with frequency f. The speed of sound is v.

The source is stationary, so its wavelength (λ_s) is unchanged: $\lambda_s = \dfrac{v}{f_s}$

The velocity of the sound waves relative to the moving observer is $(v + v_o)$

The frequency heard by the observer is:

$$f = \frac{v + v_o}{\lambda_s} = \frac{v + v_o}{\frac{v}{f_s}}$$

$$f = f_s\left(\frac{v + v_o}{v}\right)$$

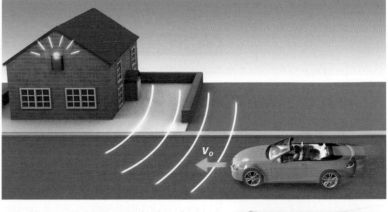

Similarly, the frequency heard by an observer moving away from a stationary source is:

$$f = f_s\left(\frac{v - v_o}{v}\right)$$

Example 3

The alarm bell in the above diagram has a frequency of 1100 Hz, and the speed of the car is 20 ms⁻¹.

a Calculate the frequency heard by the driver.

b The car changes speed after passing the alarm, and the driver hears a sound of frequency 1050Hz. Calculate the new speed of the car.

> **DON'T FORGET**
>
> The data book combines these two formulae as $f = f_s\left(\frac{v \pm v_o}{v}\right)$.

Visit http://hyperphysics.phy-astr.gsu.edu/Hbase/Sound/dopp.html for a Doppler calculator and more.

Redshift

The spectrum of light from the Sun shows absorption lines corresponding to helium.

Light from a distant star also shows absorption lines corresponding to helium, but all the lines have been moved to the left towards the red end of the star's visible spectrum.

light from the Sun

The **wavelength** of the absorption lines has **increased**, so the **frequency** of the lines has **decreased**. This Doppler effect or **redshift** provides evidence that the star is **moving away from our solar system**.

light from a distant star

LET'S THINK ABOUT THIS

1 Reminder: Moving source: use the formulae with v_s
   Moving observer: use the formulae with v_o

2 It is a common misconception that the frequency increases continuously as a fire engine approaches at speed. In fact, it is the volume or intensity of sound increasing as the fire engine moves closer that gives the impression of increasing frequency. Similarly, after the fire engine has passed, the intensity decreases but the new lower frequency heard stays constant.

INTERFERENCE – DIVISION OF AMPLITUDE

COHERENT WAVES

Two travelling waves are said to be **coherent** if they have the same **frequency and wavelength** and there is a **constant phase difference** between them.

The constant phase difference Φ between these coherent waves is:

$$\Phi = \frac{x}{\lambda} \times 2\pi$$

Destructive interference occurs if the **phase difference** between two overlapping coherent waves is π. This happens when $x = \frac{1}{2}\lambda$ and is equivalent to a **crest superimposed on a trough**.

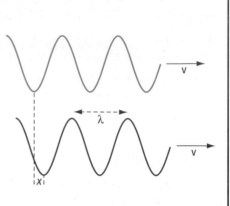

Two loudspeakers in parallel will produce coherent sound waves, as each speaker cone will move exactly in step with the other. Producing coherent light waves using two lamps in parallel is not possible due to the random nature of electron transitions and photon production in each lamp filament. Methods of producing coherent light waves will be described shortly.

Optical path difference

Interference in Higher Physics is explained in terms of path difference between two interfering waves. The concept of path difference must now be refined to include one of the waves passing through a medium other than air.

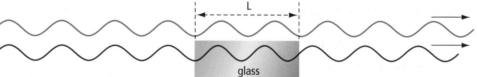

glass

The diagram represents two rays of light of equal geometric length, one travelling in **air** and the other passing through a **block of glass** of length **L**.

The **wavelength** in the glass **decreases** due to **refraction** and emerges out of phase with the wave above. It can be shown that a length **L** in a medium of refractive index **n** is equivalent to a length **nL** in air.

The **geometric** path length in the glass $= L$

The **optical** path length in the glass $= n_{glass}L$

The optical path difference (OPD) between the two rays $= n_{glass}L - n_{air}L$

$$OPD = n_{glass}L - L$$

Consider now two rays of light at near-normal incidence reflected off a glass block of refractive index n. One ray reflects off the front surface of the block while the other reflects internally off the rear surface. The glass block has a length 6 cm and refractive index 1·5.

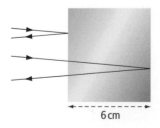

6 cm

geometric path length in the glass $= 2 \times 6\,cm = 12\,cm$ (there and back)

optical path length in the glass $= n \times$ geometric path length

The top ray doesn't travel this extra distance.

Optical path difference between the two rays $= 1\cdot5 \times (12 \times 10^{-2})$

$$= 0\cdot18\,m$$

contd

COHERENT WAVES contd

Phase changes on reflection

A light wave in air **reflecting off glass** undergoes a phase change of π.

The **leading edge** of the **incident** wave is a **crest**.

The **leading edge** of the **reflected** wave is a **trough**.

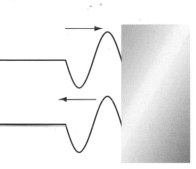

The change of phase can be seen by reflecting a single slinky pulse from a fixed end. The pulse is reflected the other way round.

A **phase change of** π occurs when light is reflected from a **higher refractive index** medium (or **optically denser** medium).

A **light wave in glass** reflecting back **into the glass** at a glass/air boundary has **no change in phase**.

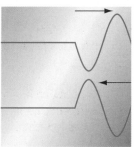

The **leading edge** of the **incident** wave is a **crest**.

The **leading edge** of a **reflected** wave is also a **crest**. There is **no phase change** at this reflection.

There is **no phase change** when light reflects from an interface with a **lower optical density** medium (**decrease in refractive index**).

Division of amplitude

Coherent sources of light are required to produce interference fringes which can be seen.

One way of doing this is to take one light wave and split it into two waves by reflection and refraction, then recombine these two waves later.

Notice that the **amplitude** of the incident wave is **greater** than the amplitudes of the **reflected** and **transmitted** waves.

An interference pattern produced by this method is called **interference by division of amplitude**. Extended light sources like the Sun or fluorescent lights can be used to produce visible interference patterns using division of amplitude.

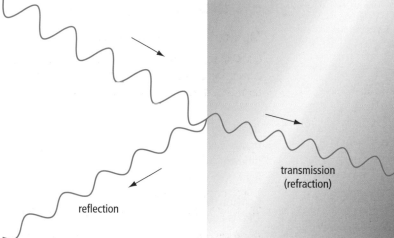

transmission
(refraction)

reflection

LET'S THINK ABOUT THIS

1   Incoherent waves also interfere but do so in a random manner. Destructive interference can only last for an instant with incoherent waves. Think of a busy swimming pool where all the water waves have different wavelengths and are out of phase with each other. No regular interference pattern is seen.

   Coherent waves, on the other hand, set up regular interference patterns which stay constant with time and are mathematically predictable.

2   A laser is one light source that does produce coherent light because each photon is emitted precisely in phase with the stimulating photon.

3   If the OPD (optical path difference) is a whole number of wavelengths, the waves will be in phase ($\Phi = 0$).

   If the OPD equals $\frac{1}{2}\lambda$, or $\frac{3}{2}\lambda$, or $\frac{5}{2}\lambda$, ... the waves will be completely out of phase ($\Phi = \pi$).

WAVE PHENOMENA

OIL FILMS

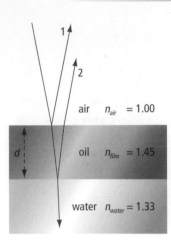

air $n_{air} = 1.00$

oil $n_{film} = 1.45$

water $n_{water} = 1.33$

A thin film of oil on a puddle of water appears multicoloured when viewed in daylight. This is an example of interference by division of amplitude.

To understand what is happening, we will consider a ray of monochromatic light falling almost vertically on the oil.

The refractive index of oil is 1·45, while water has a refractive index of 1·33.

The ray of monochromatic light is partially reflected and transmitted.

The transmitted ray reflects off the bottom surface of the oil and emerges back into the air.

It is these two **reflected** rays which interfere and cause the oil-film colours.

Ray 1 reflects with a phase change of π ($n_{oil} > n_{air}$). A phase change of π is equivalent to $\frac{1}{2}\lambda$.

Ray 2 has passed through the oil-film thickness and reflected back with no change of phase ($n_{water} < n_{oil}$).

The optical path difference (OPD) between rays 1 and 2 $= 2\,n_{oil}d + \frac{\lambda}{2}$

For destructive interference, the OPD must equal $\frac{\lambda}{2}$ or $1\frac{1}{2}\lambda$ or $2\frac{1}{2}\lambda$ etc. (i.e. an odd number of half-wavelengths).

Writing this as a formula gives: $OPD = (m+\frac{1}{2})\lambda$ where m is an integer.

Worked example

Calculate the minimum thickness of oil which will result in destructive interference of red light ($\lambda = 650\,nm$) at near-normal incidence ($n_{oil} = 1\cdot45$).

$$OPD = 2nd + \frac{\lambda}{2} = (m+\frac{1}{2})\lambda$$

$$2nd = m\lambda$$

$$2 \times 1\cdot45 \times d = 1 \times 650 \times 10^{-9} \quad (m = 1 \text{ will give the minimum value for } d)$$

$$d = 2\cdot24 \times 10^{-7}\,m \quad (0.22\,\mu m)$$

Removing this wavelength from white light will cause the reflected light from the oil on the puddle to appear coloured (green/blue). In practice, the oil film on the puddle will have many different thicknesses, so many reflected wavelengths will interfere destructively. Removing yellow light from white light will give the reflected light a purple hue.

Visit http://electron9. phys.utk.edu/ phys136d/modules/ m9/film.htm for explanation of colours on a CD and more.

Constructive interference is also possible between the two reflected rays. This happens when the OPD is equal to a whole number of wavelengths.

$$OPD = 2\,(n_{oil})d + \frac{\lambda}{2} = m\lambda, \text{ where } m = 1, 2, 3, \ldots$$

Example 1

The oil film in the worked example has a thickness of $2\cdot24 \times 10^{-7}\,m$. Show that this film will permit constructive interference of wavelengths 1300 nm (IR) and 433 nm (violet)

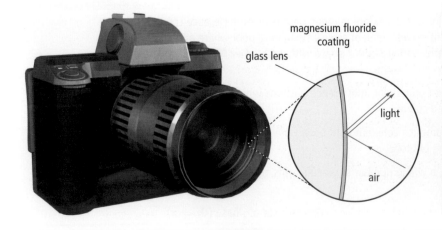

magnesium fluoride coating

glass lens

light

air

Bloomed lenses

One practical application of thin-film interference is anti-reflection coatings on camera and binocular lenses. A thin coating of magnesium fluoride is applied to the front of the lens during manufacture. A light wave reflects from the front and rear of this coating – and if these two rays interfere destructively, then the light wave will not be reflected. The resultant picture quality is improved as more light is transmitted to the film or digital sensors.

OIL FILMS contd

In more detail, consider monochromatic light falling on the coated lens at near-normal incidence.

Ray 1 has a phase change of π on reflection because $n_{coating} > n_{air}$.

Ray 2 **also** has a phase change of π on reflection because $n_{glass} > n_{coating}$.

Optical path difference between rays 1 and 2 $= 2(n_{coating})d$

For destructive interference, OPD must be $(m + \frac{1}{2}\lambda)$ with $m = 0$

(Lens coatings are applied as thinly as possible, hence $m = 0$)

$$OPD = 2(n_{coating})d = \frac{1}{2}\lambda$$
$$d = \frac{\lambda}{4n}$$

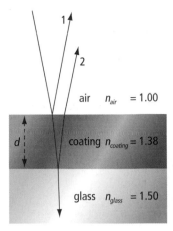

air $n_{air} = 1.00$

d coating $n_{coating} = 1.38$

glass $n_{glass} = 1.50$

Worked example

Calculate the minimum thickness of magnesium fluoride that minimises reflection of light of wavelength 570nm.

$$d = \frac{\lambda}{4n}$$
$$= \frac{570 \times 10^{-9}}{4 \times 1\cdot 38}$$
$$= 8\cdot 5 \times 10^{-8}\,\text{m}$$

There will be very little yellow light of wavelength 570nm reflected by this thickness of coating on a glass lens. There will be some reflection of colours on either side of 570nm in the visible spectrum. The relatively greater amounts of red and blue reflected light combine to give the lens a purplish hue.

700 nm	650 nm	600 nm	550 nm	500 nm	450 nm	400 nm
Red	Orange	Yellow	Green	Blue	Indigo	Violet

relative strength of reflected light

Example 2

Calculate the minimum thickness of magnesium fluoride that minimises light of frequency $5\cdot 5 \times 10^{14}$ Hz.

Example 3

A lens has a magnesium fluoride coating of thickness $1\cdot 03 \times 10^{-7}$ m. Calculate the wavelength of light for which this lens is non-reflecting.

⚙ LET'S THINK ABOUT THIS

1 Some binoculars and sunglasses have a coating with a ruby red appearance. This coating is designed to deliberately **reflect** a wavelength of this colour, and they look cool with their ruby tint. The downside is that the transmitted red colours can look a bit washed out, particularly when used on binocular lenses.

Top-end lenses have multiple layer coatings, and some internet research on this provides good background reading.

2 The effect of thin films improving light transmission was first observed by Lord Rayleigh, who noticed by chance that older tarnished lenses gave "better" light transmission than new lenses.

THIN WEDGE INTERFERENCE

A thin wedge of air between two glass plates will produce interference fringes by division of amplitude.

The incident wave is reflected from the bottom surface of the top glass plate and the top surface of the lower glass plate. The path difference between rays 1 and 2 is 2t.

Ray 2 undergoes a phase change of π, as it reflects off glass in air ($n_{glass} > n_{air}$).

Ray 1 has no phase change on reflection, as it reflects off air inside glass ($n_{air} < n_{glass}$).

The optical path difference between rays 1 and 2 $= 2t + \dfrac{\lambda}{2}$

For destructive interference, the usual condition is:

$$OPD = (m + \tfrac{1}{2})\lambda$$

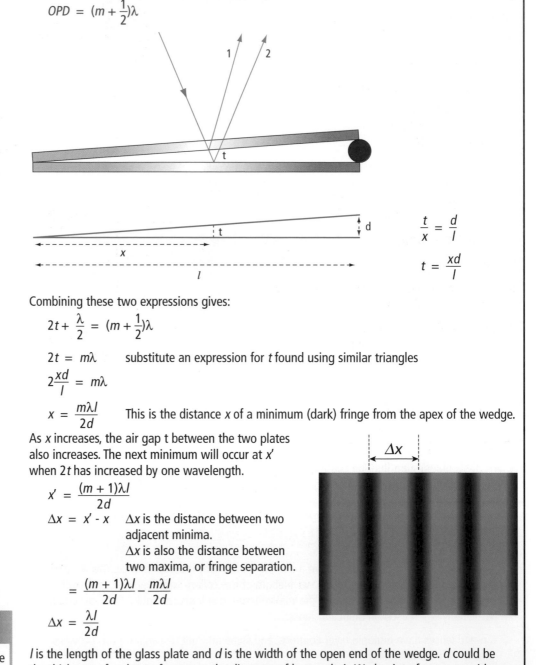

$$\frac{t}{x} = \frac{d}{l}$$

$$t = \frac{xd}{l}$$

Combining these two expressions gives:

$$2t + \frac{\lambda}{2} = (m + \tfrac{1}{2})\lambda$$

$$2t = m\lambda \qquad \text{substitute an expression for } t \text{ found using similar triangles}$$

$$2\frac{xd}{l} = m\lambda$$

$$x = \frac{m\lambda l}{2d} \qquad \text{This is the distance } x \text{ of a minimum (dark) fringe from the apex of the wedge.}$$

As x increases, the air gap t between the two plates also increases. The next minimum will occur at x' when $2t$ has increased by one wavelength.

$$x' = \frac{(m + 1)\lambda l}{2d}$$

$$\Delta x = x' - x \qquad \Delta x \text{ is the distance between two adjacent minima.}$$

$$\Delta x \text{ is also the distance between two maxima, or fringe separation.}$$

$$= \frac{(m + 1)\lambda l}{2d} - \frac{m\lambda l}{2d}$$

$$\Delta x = \frac{\lambda l}{2d}$$

DON'T FORGET

You are expected to be able to derive this expression.

l is the length of the glass plate and d is the width of the open end of the wedge. d could be the thickness of a sheet of paper or the diameter of human hair. Wedge interference provides a method of measuring the dimensions of very thin objects.

contd

THIN WEDGE INTERFERENCE contd

Measuring the diameter of a thin wire

An air wedge can be formed by placing a thin wire between two microscope slides.

A sodium light source provides the monochromatic light. A glass beam splitter reflects this light vertically down onto the air wedge.

A travelling microscope is used to view the fringes and measure their separation.

Typical results:

$$10 \text{ fringe widths} = 6\cdot5 \times 10^{-4}\,\text{m}$$

$$\text{length of glass plate} = 80\,\text{mm}$$

$$\text{wavelength of light} = 589\,\text{nm}$$

$$\Delta x = \frac{6\cdot5 \times 10^{-4}}{10} = 6\cdot5 \times 10^{-5}\,\text{m}$$

$$\Delta x = \frac{\lambda l}{2d}$$

$$d = \frac{\lambda l}{2\Delta x}$$

$$= \frac{(589 \times 10^{-9}) \times (80 \times 10^{-3})}{2 \times (6\cdot5 \times 10^{-5})}$$

$$= 3\cdot6 \times 10^{-4}\,\text{m}$$

Example 1

The wire in the experiment above is replaced by a sheet of paper of thickness $1\cdot7 \times 10^{-4}\,\text{m}$.

Calculate the fringe separation.

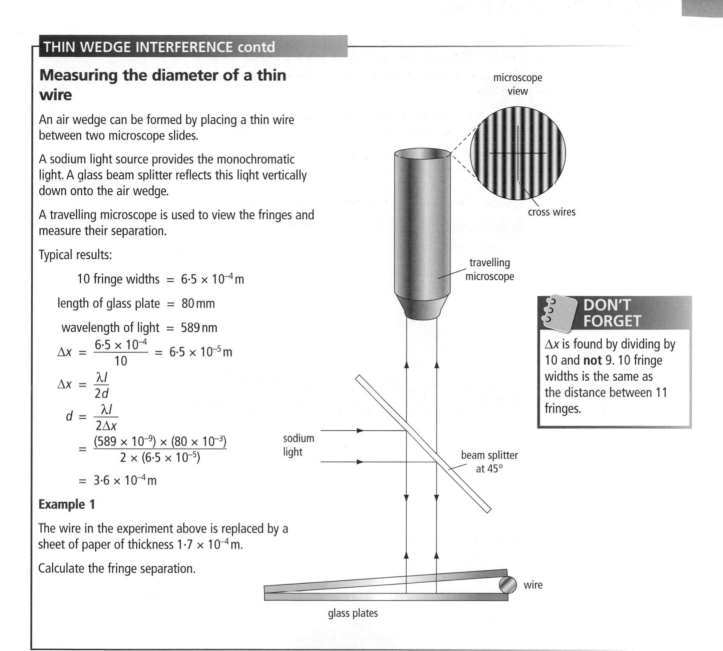

microscope view

cross wires

travelling microscope

sodium light

beam splitter at 45°

wire

glass plates

> **DON'T FORGET**
>
> Δx is found by dividing by 10 and **not** 9. 10 fringe widths is the same as the distance between 11 fringes.

LET'S THINK ABOUT THIS

If water replaced air in the wedge, the fringe separation would be $\Delta x = \dfrac{\lambda l}{2nd}$ (n = refractive index of water). Try writing out the steps to prove this relationship.

What would happen to the fringe separation if each of the following is increased?

a wavelength

b refractive index

c the angle of the wedge (try substituting $\tan\theta$ into the relationship).

INTERFERENCE – DIVISION OF WAVEFRONT

WAVEFRONTS

Circular waves spread out radially and are usually drawn as a series of concentric circles representing wavefronts travelling out in all directions.

Each wavefront joins points on the wave which are in phase and have the same wavelength and frequency. Two points on the same wavefront can become coherent sources, which can lead to an interference pattern.

This applies equally to straight plane waves. All points on the same wavefront are potential sources of coherent light.

Interference by division of wavefront takes place between two **coherent waves** which originated from the **same wavefront**. An example of interference by division of wavefront is **Young's double slit experiment**, which was studied in Higher Physics.

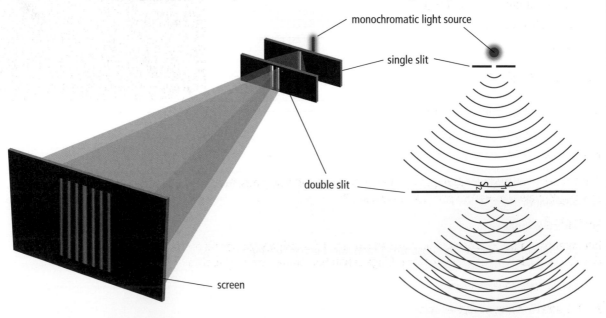

The single slit, or **collimator**, ensures that a point source of coherent wavefronts moves towards the double slit. The double slit then becomes the source of two coherent waves which diffract and interfere, producing the interference fringes on the screen.

contd

WAVEFRONTS contd

Derivation of $\Delta x = \frac{\lambda D}{d}$

Consider the region between the double slit and the screen.

The lower diagram shows a plan of the region between the zero-order and first-order maxima.

S_1 and S_2 represent the two slits.

The path difference $S_1 P - S_2 P = S_1 M = \lambda$ as P is the first maximum.

The two shaded triangles are similar, so comparing similar sides gives:

$$\frac{\Delta x}{D} = \frac{\lambda}{d}$$

$$\Delta x = \frac{\lambda D}{d}$$

The derivation assumes that angle θ is small, and so $S_2 M \cong S_1 S_2$.

Use your calculator to confirm that $\sin\theta \cong \tan\theta$ for small values of θ.

card with
two slits

Worked example

Monochromatic light incident on a double slit with a slit separation of $5\cdot7 \times 10^{-5}$ m gives this interference pattern on a screen $2\cdot5$ m away from the slits. Calculate the wavelength of the light.

$$4 \text{ fringe widths} = 80 \times 10^{-3} \text{ m}$$

$$\Delta x = 20 \times 10^{-3} \text{ m}$$

$$\Delta x = \frac{\lambda D}{d}$$

$$\lambda = \frac{\Delta x \times d}{D}$$

$$= ((20 \times 10^{-3}) \times (5\cdot7 \times 10^{-5}))/2\cdot5$$

$$= 456 \times 10^{-9} \text{ m } (456 \text{ nm})$$

80 mm

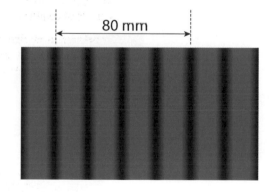

Example 1

Red light of wavelength 633 nm passes through a double slit and forms fringes 3 cm apart on a screen 2·8 m away. How far apart are the two slits?

Example 2

Interference fringes are observed on a screen when red monochromatic light is passed through a double slit. Explain what happens to the fringe separation on the screen in each of the following:

a The slit spacing is reduced.

b The screen is moved further away.

c The red light is replaced with blue light.

Visit http://surendranath.tripod.com/Applets/Optics/Slits/DoubleSlit/DblSltApplet.
html for an excellent interactive simulation of Young's slits.

POLARISATION

Unpolarised light has oscillations in **every** plane perpendicular to its direction of travel.

Polarising material (**polariser**) allows the transmission of light in one plane only due to the molecular alignment inside the material. The diagram shows transmission in the vertical plane only. (Some faint lines have been added to the polariser for purely explanatory purposes to indicate the plane of transmission.)

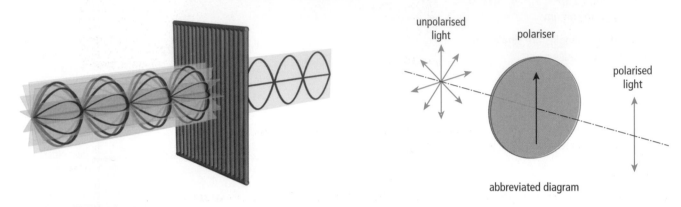

abbreviated diagram

Plane polarised light oscillates in one plane only.

A second polariser can either transmit or block this polarised light depending on its orientation. The second polariser is usually called an **analyser**; and the diagram shows blocked transmission.

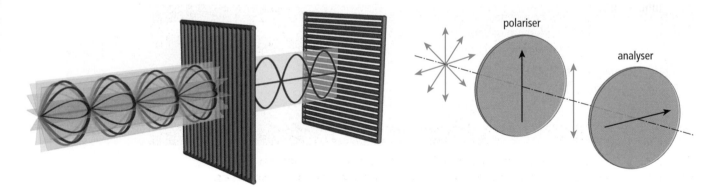

As the analyser is rotated slowly, transmission of the polarised light will increase and reach a maximum when the angle of rotation reaches 90°.

Only transverse waves can be polarised. Sound waves cannot be polarised, as they are not transverse.

Polarisation by reflection

Unpolarised light reflected off the surface of an electrical insulator like glass or water can have its reflected light fully polarised. When this happens, the angle of incidence (and reflection) is called **Brewster's angle** (i_p), and the angle between the reflected (polarised) ray and the refracted ray is 90°. It can be shown that $\tan i_p = n$ where n is the refractive index of the reflecting material.

Derivation of tan $i_p = n$

The incident ray reflects at angle i_p and refracts at angle r. The angle between the reflected and refracted rays is 90°.

The refractive index n is:

$$n = \frac{\sin i_p}{\sin r}$$

$$= \frac{\sin i_p}{\sin(180 - i_p - 90)}$$

$$= \frac{\sin i_p}{\sin(90 - i_p)}$$

$$= \frac{\sin i_p}{\cos i_p}$$

$$= \tan i_p$$

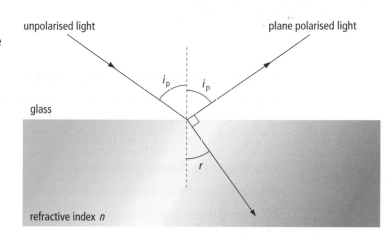

unpolarised light   plane polarised light

i_p i_p

glass

r

refractive index n

Example 1

At what angle of incidence will unpolarised light reflect off water so that the reflected light is fully polarised? The refractive index of water is 1·33.

Example 2

Unpolarised light reflects fully polarised off glass when the angle of reflection is 57°. Calculate the refractive index of the glass.

Polarising sunglasses

Sunlight, reflected off water at Brewster's angle, is plane polarised in the horizontal plane. Polarising sunglasses only transmit light oscillating in the vertical plane, so light reflected off water will not pass through.

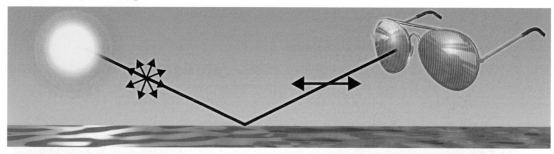

Reflections off the water, or glare, are removed, and the viewer is able to see more detail on and under the water. Photographs of water benefit from fitting a polarising filter to the camera lens.

using polarising filter

no filter

LET'S THINK ABOUT THIS

The latest 3D cinema screens use vertically and horizontally polarised light. The spectacles worn in 3D cinemas have polarising lenses, and each eye "sees" a slightly different stereo image, resulting in a 3D effect.

MEASUREMENT UNCERTAINTIES

analogue scale

$\pm \frac{1}{2}$ of the smallest division

digital scale

± 1 of the least significant digit

DON'T FORGET

There should be at least 5 measurements.

All measurements of physical quantities are subject to uncertainties; and these uncertainties can be expressed in **absolute** or **percentage** terms, e.g.

$d = 2·5 \pm 0·5$ mm $\qquad \pm 0·5$ mm is the absolute uncertainty

$d = 2·5 \pm 20\%$ mm $\qquad 0.5$ is 20% of 2·5 mm

Reading uncertainty (revision)

The reading uncertainty is a measure of how accurately an instrument's scale can be read.

The reading uncertainty on an analogue scale is $\pm \frac{1}{2}$ of the smallest division. The reading uncertainty on a digital scale is ± 1 of the least significant digit.

Voltmeter reading is $1·25 \pm 0·05$ V. Thermometer reading is $24·2 \pm 0·1$°C.

Random uncertainty (revision)

Repeating a measurement n times usually gives a spread of measurements about a mean.

Repeated measurement of the time for an event gives 3·2; 3·6; 3·3; 3·2; 3·7 seconds

$$\text{mean time} = \frac{\Sigma \; measurements}{n}$$

$$= \frac{17.0}{5}$$

$$= 3.4 \, s$$

$$\text{random uncertainty} = \frac{max \; measurement - min \; measurement}{n}$$

$$= \frac{3.7 - 3.2}{5}$$

$$= \frac{0.5}{5}$$

$$= 0.1 \, s$$

$$\text{time} = 3.4 \pm 0.1 \, s$$

Systematic uncertainty (revision)

A systematic uncertainty is an error which affects all the measurements in the same way, making them all either too high or too low, such as forgetting to zero a digital balance before a series of measurements.

Calibration uncertainty

Calibration uncertainties are given by manufacturers of scientific instruments as an indication of the accuracy of these instruments.

Typical calibration uncertainties in common lab instruments are shown in the table:

wooden metre stick:	$\pm 0·5$ mm
steel ruler:	$\pm 0·1$ mm
thermometer (liquid in glass):	$\pm 0·5$°C
analogue meter:	$\pm 2\%$ of full scale deflection
digital meter:	$\pm 0·5\%$ of the reading + 1 in the least significant digit

Worked example

The calibration uncertainty of a digital voltmeter with a reading of 2·58 V is:

$$\text{calibration uncertainty} = \pm 0·5\% \text{ of the reading} + 1 \text{ in the least significant digit}$$

$$= \pm 0·5\% \times 2.58 + 0.01$$

$$= \pm 0·013 + 0.01$$

$$= \pm 0·023 \, V$$

Total uncertainty for a measurement

Each measurement in physics should have both a reading and a calibration uncertainty. If the measurement is repeated several times, there will also be a random uncertainty. Each of these three uncertainties contributes to an overall or total uncertainty for the measurement.

$$\text{total uncertainty} = \sqrt{(reading\ uncert)^2 + (calibration\ uncert)^2 + (random\ uncert)^2}$$

Worked example

The time of 10 oscillations of a pendulum is measured 5 times using a digital stopwatch.

The results in seconds are:

17·9, 18·4, 17·4, 18·2, 17·6

$$\text{mean time} = \frac{\Sigma\ measurements}{n} = \frac{89\cdot5}{5} = 17\cdot9\,\text{s}$$

$$\text{random uncertainty} = \frac{max\ measurement - min\ measurement}{n} = \frac{18\cdot4 - 17\cdot4}{5} = \frac{1\cdot0}{5} = \pm\,0\cdot2\,\text{s}$$

reading uncertainty $= \pm\,0\cdot1\,\text{s}$

calibration uncertainty $= \pm\,0\cdot5\%$ of the reading $+ 1$ in the least significant digit

$$= \pm\,0\cdot5\% \times 17.9 + 0\cdot1$$

$$= \pm\,0\cdot09 + 0\cdot1$$

$$= \pm\,0\cdot19\,\text{s}$$

$$\text{total uncertainty in time measurement} = \pm\sqrt{(reading\ uncert)^2 + (calibration\ uncert)^2 + (random\ uncert)^2}$$

$$= \pm\sqrt{(0\cdot1)^2 + (0\cdot19)^2 + (0\cdot2)^2}$$

$$= \pm\sqrt{0\cdot086}$$

$$= \pm\,0\cdot29$$

$$= \pm\,0\cdot3\,\text{s}$$

$$\text{time of 10 oscillations} = 17\cdot9 \pm 0\cdot3\,\text{s}$$

Dominant uncertainty

Uncertainties less than $\frac{1}{3}$ of the dominant uncertainty can be ignored. This often simplifies the procedure to find the uncertainty associated with a measurement.

Worked example

The uncertainty in the measurement of a temperature provides the following:

reading uncertainty $= \pm\,0\cdot5\,°\text{C}$

calibration uncertainty $= \pm\,0\cdot5\,°\text{C}$

random uncertainty $= \pm\,0\cdot1\,°\text{C}$

The random uncertainty is less than $\frac{1}{3}$ of the other two uncertainties so can be ignored.

$$\text{total uncertainty in temperature} = \pm\sqrt{(reading\ uncert)^2 + (calibration\ uncert)^2}$$

$$= \pm\sqrt{(0\cdot5)^2 + (0\cdot5)^2}$$

$$= \pm\,0\cdot7\,°\text{C}$$

Taking several measurements of the temperature will reduce the random error, and in this example it can be ignored compared to reading and calibration uncertainties. Taking **even more** temperature measurements will have **no effect** on the total uncertainty in temperature, as the other uncertainties are dominant **in this example**.

UNCERTAINTIES

Example 1

A current is measured using a digital ammeter. Calculate the total uncertainty of the current measurement with this data:

$$\text{reading uncertainty} = \pm 0.1 \, \text{mA}$$

$$\text{calibration uncertainty} = \pm 0.11 \, \text{mA}$$

$$\text{random uncertainty} = \pm 0.4 \, \text{mA}$$

UNCERTAINTY IN A PRODUCT OR QUOTIENT OF QUANTITIES

DON'T FORGET

Fractional uncertainties can replace % uncertainties here. Most students prefer the % uncertainty approach.

Consider the relationships $X = Y \times Z$ or $X = Y/Z$. The percentage uncertainty in X ($\%\Delta X$) is found using:

$$\%\Delta X = \pm \sqrt{(\% \text{ uncert in } Y)^2 + (\% \text{ uncert in } Z)^2}$$

Worked example

The moment of inertia of an object about an axis and its angular velocity about the same axis are:

$$I = 9.9 \times 10^{-4} \pm 4 \times 10^{-5} \, \text{kgm}^2$$

$$\omega = 4.3 \pm 0.3 \, \text{rads}^{-1}$$

Calculate the angular momentum of the object and the absolute uncertainty in the calculated value.

$$L = I\omega = (9.9 \times 10^{-4}) \times 4.3 = 4.3 \times 10^{-3} \, \text{kgm}^2\text{s}^{-1}$$

$$\% \text{ uncertainty in } I = \frac{4 \times 10^{-5}}{9.9 \times 10^{-4}} \times 100 = 4\%$$

$$\% \text{ uncertainty in } \omega = \frac{0.3}{4.3} \times 100 = 7\%$$

$$\% \text{ uncertainty in } L = \pm \sqrt{(\% \text{ uncert in } I)^2 + (\% \text{un cert in } \omega)^2}$$

$$= \pm \sqrt{(4)^2 + (7)^2}$$

$$= \pm 8\%$$

$$8\% \text{ of } 4.3 \times 10^{-3} = 3 \times 10^{-4}$$

$$L = 4.3 \times 10^{-3} \pm 3 \times 10^{-4} \, \text{kgm}^2\text{s}^{-1}$$

Example 1

Two parallel plates are separated by a distance d and have a voltage V across them. Measurements of these are:

$$V = 1250 \pm 100 \, \text{V}$$

$$d = 2.50 \times 10^{-2} \pm 1 \times 10^{-3} \, \text{m}$$

DON'T FORGET

Any uncertainty less than $\frac{1}{3}$ of the dominant uncertainty can be ignored.

Calculate the electric field strength E between the plates and the associated percentage and absolute uncertainties in E.

If more than two quantities are combined by multiplication or division, then the formula is extended:

$$\%\Delta W = \pm \sqrt{(\% \text{ uncert in } X)^2 + (\% \text{ uncert in } Y)^2 + (\% \text{ uncert in } Z)^2 + \ldots}$$

UNCERTAINTY IN A QUANTITY RAISED TO A POWER

If $P = A^n \Rightarrow$ % uncertainty in $P = n \times$ % uncertainty in A

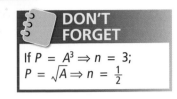

DON'T FORGET

If $P = A^3 \Rightarrow n = 3$;
$P = \sqrt{A} \Rightarrow n = \frac{1}{2}$

Worked example

$\omega = 51 \pm 2\,\text{rads}^{-1}$. Calculate ω^2 and the uncertainty in ω^2.

$\omega^2 = 51^2 = 2\cdot6 \times 10^3\,rad^2s^{-2}$

% uncertainty in $\omega = \dfrac{2}{51} \times 100 = 4\%$

% uncertainty in $\omega^2 = 2 \times 4 = 8\%$

$\omega^2 = 2\cdot6 \times 10^3 \pm 8\%$

$\omega^2 = 2\cdot6 \times 10^3 \pm 2 \times 10^2\,rad^2s^{-2}$

Example 2

Calculate the energy stored in an inductor and the uncertainty in energy using this information:

$L = 2\cdot5 \pm 0\cdot2\,\text{H}$

$I = 8\cdot3 \pm 0\cdot5\,\text{mA}$

UNCERTAINTY IN A SUM OR DIFFERENCE OF QUANTITIES

The uncertainty in a sum or difference of quantities uses absolute uncertainties and not percentage uncertainties.

If $X = Y + Z,$ or $X = Y - Z$

absolute $\Delta X = \pm \sqrt{(absolute\ \Delta Y)^2 + (absolute\ \Delta Z)^2}$

Worked example

Use this data to calculate the value and uncertainty in $v + v_s$ (used in Doppler effect)

$v = 340 \pm 5\,\text{ms}^{-1}$

$v_s = 30 \pm 3\,\text{ms}^{-1}$

$v + v_s = 340 + 30 = 370\,\text{ms}^{-1}$

uncertainty in $v + v_s = \pm\sqrt{3^2 + 5^2} = \pm 6\,\text{ms}^{-1}$

$v + v_s = 370 \pm 6\,\text{ms}^{-1}$

Example 3

An analyser in a polarisation experiment is rotated relative to a fixed protractor:

position 1 $= 55 \pm 1°$

position 2 $= 74 \pm 1°$

Calculate the angle of rotation and its uncertainty.

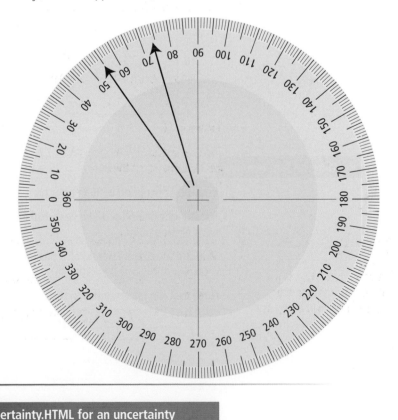

Visit http://web.mst.edu/~gbert/JAVA/uncertainty.HTML for an uncertainty calculator.

GRAPHICAL ANALYSIS OF UNCERTAINTIES

Uncertainty in the position of a point on a graph is shown using error bars. The lengths of the horizontal and vertical bars is a measure of the uncertainty in each coordinate. The following table of results from a current balance experiment contains uncertainties, and the corresponding graph illustrates how error bars are drawn.

Current/mA	0	100 ± 10	200 ± 10	300 ± 10	400 ± 10	500 ± 10	600 ± 10

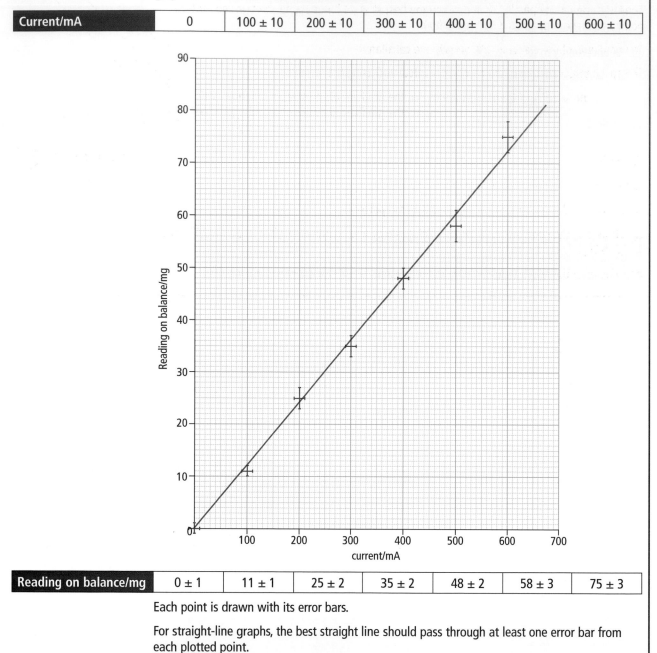

Reading on balance/mg	0 ± 1	11 ± 1	25 ± 2	35 ± 2	48 ± 2	58 ± 3	75 ± 3

Each point is drawn with its error bars.

For straight-line graphs, the best straight line should pass through at least one error bar from each plotted point.

Straight-line graphs should also pass through the centroid (or average x and y point). The centroid of this graph is (300, 36).

Note that the uncertainties in the balance reading increase, so the size of the balance-reading error bars increases to reflect this.

contd

GRAPHICAL ANALYSIS OF UNCERTAINTIES contd

Physicists often use the gradient of a straight-line graph in further calculations. One way of finding the uncertainty in the gradient is to draw a parallelogram around the best straight line.

Draw a line parallel to and above the best straight line, passing through the furthest **point** (not error bar) from the best line (red line).

Similarly, draw a line parallel to and below the best straight line, passing through the furthest point below the best straight line (blue line).

The gradients of the diagonals AC and DB are calculated.

The absolute uncertainty in the gradient is calculated using the formula:

$$\Delta m = \frac{m_1 - m_2}{2\sqrt{(n-2)}}$$

where m_1 = the gradient of the steeper diagonal (DB)
m_2 = the gradient of the less steep diagonal (AC)
n = the number of points on the graph

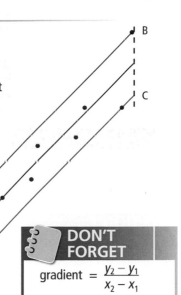

DON'T FORGET

$$\text{gradient} = \frac{y_2 - y_1}{x_2 - x_1}$$

Example 4

The following questions refer to the graph opposite.

a Show that the gradient of the graph is 0·12 mg per *mA*.

b Show that the coordinates of the parallelogram of uncertainty are (0,3), (0,–3), (600,75) and (600,70).

c Show that the absolute uncertainty in the gradient is ± 4·5 × 10⁻³ mg per *mA*.

INVESTIGATION REPORT

You are required to prepare a report after completing your AH Physics investigation. This report is marked externally by an SQA marker and is worth a total of 25 marks.

Guidance covering all aspects of the investigation is given in the Candidate Information leaflet produced by SQA. Make sure you have a copy of this (it can be downloaded from the SQA website). Some **additional** advice on the preparation of the final investigation report is included here.

INTRODUCTION

DON'T FORGET

The summary is called an **abstract** in professional scientific reports.

Summary A summary of the **purpose** and **findings** of the investigation must be included at the start of the report. Students regularly fail to include the findings in their summary, e.g. *measuring refractive index of water by several methods* but not listing what the results were; or *comparing methods of measuring the speed of sound* and not summarising the results of the comparison. **(1 mark)**

Underlying physics Each experiment described should involve physics relationships or formulae, and the report should demonstrate an understanding of the physics behind these formulae and the way they are used in the investigation. Simply quoting the formula for the surface tension of a liquid without an attempt at explanation will not get 3 marks. Writing big chunks of theory from a reference source does not necessarily demonstrate understanding. **(3 marks)**

PROCEDURES

Diagrams and/or descriptions of the apparatus Unlabelled or unclear diagrams will lose marks. Digital pictures should be labelled and care taken that the final version in the report does not lose clarity. Photographs of circuits should be supplemented by a circuit diagram. **(2 marks)**

Use of apparatus to obtain experimental data A clear description of the experimental method is required. **(2 marks)**

Level of demand The investigation must be at an appropriate level and not an extended Standard Grade investigation.

Three to four related experiments should be attempted. Coursework from AH or H Physics should be limited to an initial experiment only. **(2 marks)**

RESULTS

Data sufficient/relevant Show all readings taken during the course of the investigation. **(1 mark)**

Uncertainties Each measurement should have calibration, scale reading and random uncertainties where appropriate, and these combined for a total uncertainty. Show one example only of a calculation of the total uncertainty for a measurement.

A quantity which is calculated using a formula should also have an absolute or percentage uncertainty found by combining the uncertainties as described earlier.

The absolute uncertainty of the final result should have **one significant figure only**, e.g.

$$g = 9.6 \pm 0.5 \, \text{ms}^{-2}$$

$$\text{or} \quad g = 9.73 \pm 4 \times 10^{-2} \, \text{ms}^{-2}$$

RESULTS contd

Visit the Bright Red website to view the uncertainties booklet produced for AH Physics.

(3 marks)

Analysis of data The investigation report must show how the data is used to calculate numerical values and to draw graphs. If a spreadsheet package is used to produce a graph, make sure the graph is not too small, grid lines are included and the graph drawn is not "dot to dot".

(2 marks)

DISCUSSION

Conclusion Your report should include an overall conclusion which is valid for the experimental results.

(1 mark)

Evaluation of the experimental procedures Include an evaluation after each experiment discussing accuracy of your measurements, control and range of variables, limitations of equipment and sources of error and uncertainty.

Try to avoid the temptation to say that "better" results would be obtained with "more accurate equipment".

(3 marks)

One of your graphs may be a straight line but just misses the origin. Comment on this and think which systematic error may be responsible.

An experimental result for the acceleration due to gravity of $g = 9.4 \pm 0.2 \, \text{ms}^{-2}$ should have an attempt at explaining why it is less than the accepted value and also why the uncertainty limits do not extend to the accepted value.

Stating and discussing matters like these demonstrates good physics evaluation.

Evaluation of the investigation as a whole Have this as a separate heading at the end of your report. Avoid simply restating discussion points from the individual evaluations. Describe something that went wrong and how it was resolved. Think of further work that could be done.

(2 marks)

PRESENTATION

Title, contents, pages numbered An easy mark for most reports (but not all!). **(1 mark)**

Clear, concise, readable Redrafting your work helps to make the report more readable.

(1 mark)

References References should be cited in the text of the report using "(Ref 1)" beside the piece of information or formula you wish to refer to. At the end of the report, a references page should give the full name of the reference book **and the page number**.

The "underlying physics" pages near the start of the report should have some references.

Many reports still fail to score this mark either because the reference is not cited in the text **or** because page numbers have been omitted.

(1 mark)

ANSWERS

ANSWERS

Mechanics
pp. 4–5

Example 1

a $v = 40t - 120t^3$

b $a = 40 - 360t^2$

c 15 m

d zero

e 40 ms⁻²

f 0 s and 0·58 s

Example 2

a $v = 3t^2 + 12t$

b $a = 6t + 12$

c 576 ms⁻¹

d 10·4 km

Let's think about this

8°8 s

pp. 6–7

Example 1

1·766 × 10⁻²⁷ kg

Example 2

1·9 × 10⁸ ms⁻¹

Example 3

2·35 × 10⁻¹⁰ J

Example 4

1·64 × 10⁸ ms⁻¹

pp. 8–9

Example 1

1·57 rads⁻¹

Example 2

7·3 × 10⁻⁵ rads⁻¹

Example 3

24·6 hours; 1·88 years

Example 4

2·3 × 10³ radians

Example 5

a 465 ms⁻¹

b 260 ms⁻¹

Example 6

2·1 × 10⁻⁵ ms⁻¹

Example 7

3 × 10⁴ ms⁻¹ or 30 kms⁻¹

pp. 10–11

Example 1

A: 0·7 rads⁻²; B: 3·6 rads⁻²

Example 2

14 s

Example 3

3·4 revolutions

Example 4

1·6 s

Example 5

a 40 rads⁻²; −30 rads⁻²

b 1020 rad

c 162

pp. 12–13

Example 1

a 0·34 rads⁻²

b 1·4 ms⁻²

c **i** 27·7 ms⁻²

  ii 59 ms⁻²

Example 2

71 N

Example 3

900 rpm

pp. 14–15

Example 1

2·0 × 10⁻² kgm²

Example 2

a 7·3 × 10⁻⁵ rads⁻¹

b 9·8 × 10³⁷ kgm²

c 2·6 × 10²⁹ J

Example 3

c greater; more mass at a greater distance from the axis

Example 4

0·33 J

Example 5

1·8 × 10⁻² kgm²

Example 6

0·15 m

Let's think about this

1 I will be less as more mass is concentrated nearer the axis. E_k will be less as well, as I is less.

pp. 16–17

Example 1

93 N

Example 2

15 Nm

Example 3

a 0·11 Nm

b 0·20 Nm

pp. 18–19

Example 1

a 10 rads⁻¹

b 0·35 J

Example 2

4·5 × 10⁻³ kgm²

Example 4

a 0·24 kgm² rads⁻¹

b 6·9 rads⁻¹

Let's think about this

I_{earth} increases, as more mass further from axis; ω decreases

pp. 21–22

Example 1

1·2 m

Example 2

min 8·8 × 10¹⁷ N; max 1·9 × 10¹⁸ N

Example 3

$r/ \times 10^6$ m	6.4	8	10	15	20
g/Nkg⁻¹	9.8	6·3	4	1·8	1

Example 4

1·7 Nkg⁻¹

Let's think about this

2 midday; you are closer to the Sun

pp. 22–23

Example 1

−2·6 × 10⁶ Jkg⁻¹

Example 2

420 km

Example 3

2·4 × 10³ ms⁻¹

Example 4

1·09 × 10⁴ ms⁻¹

Example 5

177 ms⁻¹

pp. 24–25

Example 1

115 km

Example 2

36 000 km

Example 4

5·7 × 10²⁶ kg

Example 5

348 km

Example 7

3·0 × 10⁴ ms⁻¹

Let's think about this

365·2 days

pp. 26–27

Example 1

a 7·2 Hz

b 380 N

c 65·9 ms⁻¹

pp. 28–29

Example 2

24·8 cm

pp. 30–31

Example 1

7·3 Hz

Example 3

a 0·1 J

b 0·4 J

c 0·3 J

d 1·5 ms^{-1}

e 1·8 ms^{-1}

f $y = 0$

pp. 32–33

Example 1

$4·9 \times 10^{-11}$ m

Let's think about this

$\dfrac{1}{\sqrt{2}}$

pp. 34–35

Example 1

$1·06 \times 10^{-34}$ kg m^2 s^{-1}

Example 2

$n = 4$; fourth orbit

Electrical phenomena

pp. 36–37

Example 1

a 4·5 N to the left

b $2·4 \times 10^{-4}$ N to the left

Example 2

4·1 N to the left

Example 3

1·6 N at 310°

pp. 38–39

Example 1

a 5×10^{-5} NC^{-1}

b $2·8 \times 10^{-5}$ NC^{-1}

Example 2

0·7 m

Example 3

$4·7 \times 10^7$ NC^{-1} to the right

pp. 42–43

Example 1

$-0·72$ V

Example 2

$X -3·4 \times 10^4$ V, $Y\, 1·8 \times 10^4$ V

Example 3

$1·2 \times 10^{-19}$ J

pp. 44–45

Example 1

0·08 m

Example 2

9·3 cm

pp. 46–47

Example 1

$1·8 \times 10^{-12}$ m

Example 2

$4·8 \times 10^5$ m s^{-1}

Example 3

platinum

pp. 48–49

Example 1

a $7·5 \times 10^{-5}$ N into the page

b $6·3 \times 10^{-4}$ N upwards towards top of the page

pp. 52–53

Example 1

2·5 μT

Example 2

13 A

Example 3

1·1 mm

Example 4

3·9 A, same direction

Let's think about this

2 the speed of light

pp. 62–63

Example 1

0·72 J

Example 2

80 mA

pp. 68–69

Example 1

increases by $\sqrt{2}$

Example 2

$\dfrac{1}{4}$ of its original value

Example 3

$2·5 \times 10^{-3}$ m, 7·2 Hz, 9·0 m, 65 ms^{-1} to the left

Example 4

$y = 0·04 \sin(60\pi t - 80\pi x)$

pp. 70–71

Example 1

322 ms^{-1}

Example 2

1·4 kHz

pp. 72–73

Example 1

approaching, 13·6 ms^{-1}

Example 2

1740 Hz

Example 3

a 1165 Hz

b 15·5 ms^{-1}

pp. 76–77

Example 2

99 nm

Example 3

569 nm

pp. 78–79

Example 1

$1·4 \times 10^{-4}$ m

Let's think about this

a increases

b decreases

c decreases

pp. 80–81

Example 1

$5·9 \times 10^{-5}$ m

Example 2

a Δx increases; as $d\downarrow$, $\Delta x\uparrow$

b Δx increases; $D\uparrow$, $\Delta x\uparrow$

c $\lambda_{blue} < \lambda_{red}$; $\lambda\downarrow$ $\Delta x\downarrow$; Δx decreases

pp. 82–83

Example 1

53°

Example 2

1·54

pp. 86–87

Example 1

± 0·4 mA

Example 2

$5·0 \times 10^4$ Vm^{-1}; ± 9%; ± $4·5 \times 10^3$ Vm^{-1}

Example 3

$8·6 \times 10^{-5}$ ± $1·2 \times 10^{-5}$ J

Example 4

19 ± 1·4°

INDEX